PRACTICAL SHEET METAL WORK AND DEMONSTRATED PATTERNS

A COMPREHENSIVE TREATISE IN SEVERAL VOLUMES ON SHOP AND OUTSIDE PRACTICE AND PATTERN DRAFTING

VOLUME XI
AUTOMOBILES AND SHEET METAL BOATS

BY
WILLIAM NEUBECKER

COMPILED FROM THE
METAL WORKER
PLUMBER AND STEAM FITTER

DAVID WILLIAMS COMPANY
NEW YORK
1912

Practical Sheet Metal Work
and Demonstrated Patterns

Automobiles and Sheet Metal Boats

by William Neubecker

Originally published by
David Williams Company
New York

Reprinted by
Lindsay Publications Inc
Bradley IL 60915

ISBN 1-55918-256-3

1 2 3 4 5 6 7 8 9 0

2001

WARNING

Remember that the materials and methods described here are from another era. Workers were less safety conscious then, and some methods may be downright dangerous. Be careful! Use good solid judgement in your work, and think ahead. Lindsay Publications Inc. has not tested these methods and materials and does not endorse them. Our job is merely to pass along to you information from another era. Safety is your responsibility.

PREFACE

THE following series of articles are devoted to the sheet metal features of automobile construction, and give an authoritative presentation of the state of the art to-day. Those already engaged in the work will find them interesting as giving an insight of the methods pursued in factories other than those with which they are identified. They will interest those not engaged in the work, in the opportunity they afford to learn the degree of development of what is relatively a new branch of the sheet metal industry. They will prove of value to still another branch of sheet metal workers; those operating jobbing shops where demands for repair parts or renewal parts may be made.

The articles aim to present short geometrical rules for developing the various patterns required in the sheet metal portion of the automobile's makeup. Care has been taken to show how and where flanges must be added to the patterns for nailing to the ash frame of the automobile, for riveting to the aluminum trimmings of the machine, for bolting to the channel of the chassis. Where necessary, photographs and sketches are used to make clear the various methods of constructing the joints and to indicate the way the joints are riveted and soldered. Sketches and descriptions of the various hand tools employed and the foot and power presses are also given. The figures illustrate the various sheet metal parts of the automobile, and give the proper name for each part, so that it is possible for the reader to intelligently follow the matter descriptive of that part. So far as possible the treatment of the articles involves the consideration of each part in the regular order in which it is taken up in the actual erection of the automobile in the shop.

The section on sheet metal boats is compiled from the columns of METAL WORKER and clearly illustrates the layout and construction of canoes, rowboats and launches.

WILLIAM NEUBECKER.

July 25, 1912.

CONTENTS

PAGE

Constructing the Bodies and Panels

The Various Sheet Metal Parts of the Automobile 2
The Scale Drawing 3
Patterns for Typical Automobile Body 5
Cutting and Punching the Body 8
Fastening Body to Chassis 9
Patterns for Front and Rear Panels 9
Constructing the Panels and Body 10
Soldering and Smoothing the Riveted Joints 12
Developing Patterns for Automobile Bodies with Different Profiles . . . 13
Various Types of Radiators 14

Patterns for Bodies and Dash Hoods

Patterns for Body and Dash Hood Combined in an Automobile whose Sides are Straight, the Hood being on a Horizontal Plane 15
Bending and Slotting the Tubing for Dash Hoods 16
Pattern for Body and Dash Hood Combined for Touring Car, the Sides of the Body being Straight and the Dash Hood being Developed by the Parallel Line Method 17
Pattern for Body and Dash Hood Combined for a Raceabout, the Sides of tne Body being Straight, the Dash Hood being Developed by Parallel Lines . . . 19
Pattern for Body and Dash Hood of a Touring Car, whose Sides are Straight, the Outline of the Dash Hood having a Double Curve and being Developed by Parallel Lines 21
Pattern for Body and Dash Hood of a Torpedo Type of Automobile having Straight Sides, the Hood being Developed by Parallel Lines 22
Pattern for Body and Dash Hood of a Tourabout with Straight Sides, the Dash Hood requiring Triangulation in its Solution 24
Pattern for Body and Dash Hood of a Roadster, the Sides of the Body being Straight with Tapering Dash Hood, the Hood being Developed by Radial Lines . 27
Pattern for a Different Outline of Conical Dash Hood 29
Pattern for Body and Dash Hood Combined of Semi-Racer, whose Sides are Straight and whose Hood is Developed by Triangulation 31
Pattern for Body and Dash Hood of a Roadster Type, whose Body Sides are Straight and whose Hood is Developed by Triangulation 33
Pattern for Body and Dash Hood Combined of a Runabout whose Sides are Straight, the Dash Hood being Developed by Triangulation 36
Pattern for Body and Dash Hood of a Roadster having Straight Sides, the Hood being Developed by Parallel Lines 37

CONTENTS (*Continued*)

PAGE

Pattern for Body and Dash Hood of a Runabout, whose sides are curved, the Hood having a Compound Curve being Developed by Triangulation 39

Pattern for Dash Hood and Front Part of Body in a Gunboat Type of Automobile, the Body being Curved and Hood being Developed by Triangulation . . 42

Pattern for Dash Hood and Front Part of Body in a Torpedo Type of Automobile, the Sides of the Body being Straight, and the Dash Hood being Developed by Triangulation 45

Pattern for Tapering Dash Hood Combined with Body in a Racer Car, the Dash Hood joining a Motor Hood of Peculiar Shape, requiring Triangulation in its Solution 48

Method of Fastening Sheet Metal Dash Hoods to Wooden or Sheet Metal Dash Boards 51

Developing and Constructing Seats and Doors

Developing the Patterns and Constructing Sheet Metal Seats 53

Pattern for Two Passenger Seat 54

Constructing the Seat 57

Pattern for Individual Seats 60

Pattern for Single Rumble Seat 63

Pattern for Single Rumble Seat, without Flare 65

Pattern for a Double Rumble Seat 66

Pattern for a Racer Seat 68

Pattern for a Surrey Seat 70

Pattern for a Motor Truck Seat 71

Constructing the Sheet Metal Door 72

Motor Hoods

Constructing and Developing Motor Hoods 74

Pattern for Tapering Square Molded Motor Hood 76

Pattern for Flaring Piece in a Semi-Circular Motor Hood 77

Patterns for Irregular Shaped Motor Hood 78

Patterns for French Type of Motor Hood 80

Pattern for French Motor Hood, without Chamfer 83

Guards and Fenders

Pattern and Construction of Front Wheel Fenders and Splash Guards . . 86

Pattern for Fender joining Channel Frame and Running Board Mud Guard . 87

Constructing the Mud Guard 90

Pattern for Projecting Rim or Lip of Mud Guard 92

Pattern for Mud Guard joining Running Board Only 93

Pattern for Mud Guard having a Double Curve 95

Pattern for Mud Guard with Fender, Curved both in Plan and Elevation, and Intersected by Running Board Splash Guard 97

CONTENTS (*Continued*)

PAGE

Pattern for a Flaring Fender 101
Pattern for Flaring Twisted Fender 102
Hind Wheel Fenders 104

Separate Dash Hoods

Pattern for Separate Dash Hood Developed by Parallel Lines . . . 104
Pattern for Separate Dash Hood Developed by Triangulation 105
Pattern for Separate Dash Hood joining Door Frame 108
Pattern for Separate Dash Hood on a Torpedo Model 110

Tanks, Funnels and Accessories

Constructing Gasoline Tanks 112
The Oil Tank 113
Fly Wheel Hood 114
Mud Apron or Pan 114
The Muffler 114
Types of Funnels Used 114
Pattern for Center Funnel Oval to Round 115
Pattern for Funnel, Perpendicular at one end 116
Patterns for a Rectangular Funnel 118
Patterns for Combination Measure and Funnel 120

BOAT PATTERNS

Two-Piece Sheet Metal Practice Boat 123
Constructing Sheet Metal Boats on Skeleton Frames 127
Building Metal Boats by Means of Models 130
Sheet Metal Canoe 132
Building Sheet Metal Boats from Gores or Strips 135
Patterns for a Sheet Metal Boat Developed by Triangulation 137

AUTOMOBILE SHEET METAL PATTERNS

Practical Article on Obtaining Patterns for Different Parts of the Sheet Metal Work in the Manufacture of Automobiles, Together with the Various Methods of Constructing the Seams and Joints

The automobile industry today not only gives employment to thousands of machinists, carpenters, enamelers, upholsterers, varnishers, engineers, etc., but also to thousands of sheet metal workers in the various automobile plants where the

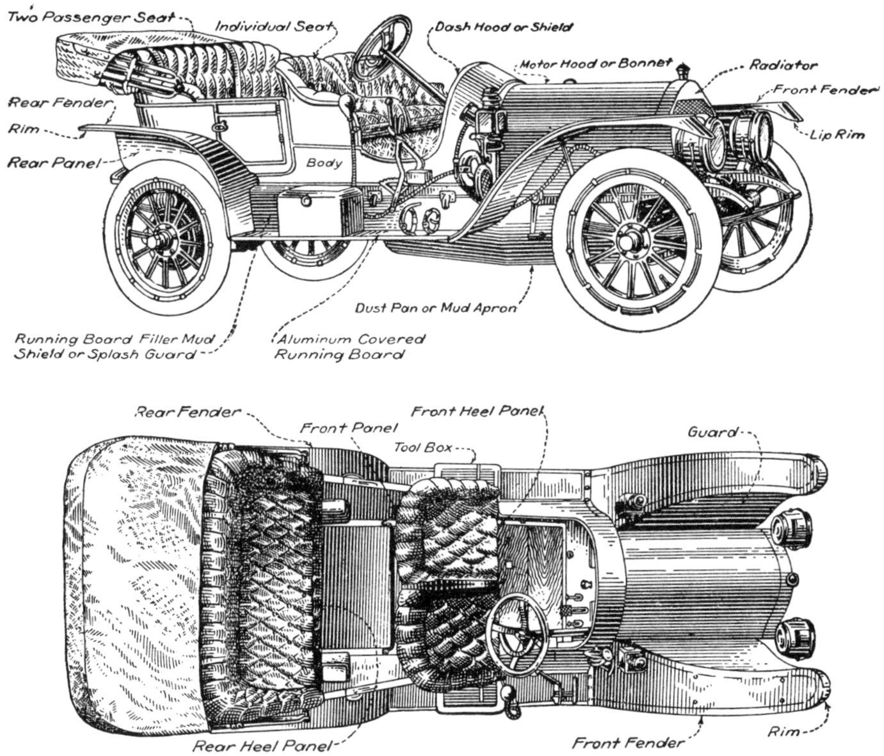

Fig. 1. Top and Side View of a Typical Finished Automobile

various parts of the body of the automobile are made from sheet metal. One would be surprised on entering the sheet metal department of an up-to-date automobile plant to see the progress made in the manufacture of the sheet metal bodies, heel panels, single and double seats, doors, fenders, fillers, French hoods, tapering hoods, dash hoods, aprons, mufflers, fly-wheel covers, tanks, dust pans, running board fillers, mud aprons, etc., all made from either No. 20 black sheet steel, aluminum coated sheets or pure sheet aluminum.

While some years ago the bodies of the automobile were made of wood, it was found that they warped, shrunk or cracked and an accident meant a splintered part or possibly a fire. To day every concern manufacturing high grade cars has accepted the metal body as the ideal equipment and one of the most prominent selling features. The metal body and its parts lend themselves to a more permanent finish in painting than the wooden body, because they offer a non-porous and smooth surface to the painters' brush. Cast aluminum trimmings are used in connection with the metal body and give rigidity and prevent buckling.

Some claim black sheets should not be used, owing to their extreme susceptibility to rust, resulting in the paint scaling and in further corrosion. While aluminum coated sheets or pure sheet aluminum naturally makes a better job, still, if the various parts of the sheet steel bodies are cleaned with the sand blast, this will remove all scale and rust, and if the metal is immediately painted and then enameled, corrosion is then minimized. Whatever metal material is used for the body and its parts, this, together with the aluminum trimmings results in an ideal automobile body that forms a unit for itself, and is absolutely independent of the chassis or lower frame of the car and of all the stresses and strains borne by it. The rigidity of the aluminum trimmings and inner ash framings to the metal body eliminate all buckling tendencies.

The Various Sheet Metal Parts

So that the readers may be familiar with the various shop terms used in sheet metal automobile construction, the various names of the sheet metal parts in the automobile are shown in Fig. 1, which gives a side and top view of a typical finished automobile. Thus we have the rear fender, rim and splash guard, which when joined together, constitutes the rear mud guard. There is also the front fender, rim and splash guard, which constitutes the front mud guard. Then comes the body, to which is attached the front heel panel, placed below the front of the front seat and the rear heel panel, placed below the front of the rear seat. Then comes the front panel; which is placed in the back of the front seat and the rear panel which forms the rear part of the body. The dash hood or shield and the radiator is also shown, as well as the motor hood or bonnet between these two. In the side as well as the top views, there are shown the two-passenger seat and the individual seat. The door as well as the tool box are indicated in the side view. The dust pan or mud apron protects the lower works from flying gravel, mud, dust and splashing water. The running board is sometimes cast entirely from aluminum or made of wood and covered with sheet aluminum. The space between the running board and body of the car is protected by the running board filler or mud shield, which keeps down the dust, mud, etc.

Other parts belonging to the automobile, which are made from sheet metal, but are not shown, are the fly-wheel cover, the oil tank, the gasoline tank and the muffler, which is used to deaden the noise of the escaping gases. The oil tank is made of IXX tin or 16 oz. copper, while the gasoline tank is often made of 20 oz. cold rolled copper or No. 20 galvanized iron. All the rest of the sheet metal work just named is made of No. 20 black sheet steel, sheet aluminum or aluminum covered sheets.

Knowing the various shop terms the practical construction is now in order. A very important part of an automobile plant is the drafting room in which the various designs are laid out, scale drawings and blue prints prepared and from which the various detail drawings are constructed, ready for development in sheet metal by the shop foreman.

The Scale Drawing

A scale drawing is as important in the building of an automobile as a plan in the building of a sheet metal front or cornice. Upon the designer depends the graceful outlines and proportions of the automobile. Upon his advice and under his care the draftsman prepares the various scale drawings of the automobile, which are usually drawn 1½ in. to the foot. Thus the iron worker, the machinist, the carpenter, the sheet metal worker, etc., all receive their blue prints, giving full size measurements for the various parts required in the different departments. Fig. 2 shows a typical scale drawing with measurements of the sheet metal work required in the building of the various parts of the body. A general description of the various parts may not be out of place.

In the half rear elevation A shows a section of the channel of the chassis proper, upon which the lower part of the body B C in the side elevation rests, the body being fastened to the channel by means of an ash frame, a section of which is shown by D in the half front elevation. In the half rear elevation E F shows the curve of the lower part of the body, while r t in the side elevation shows the curve of the body shown in the half rear elevation. This miter intersection between the rear and side of the body is shown by r' t' in the half bottom view. The letter

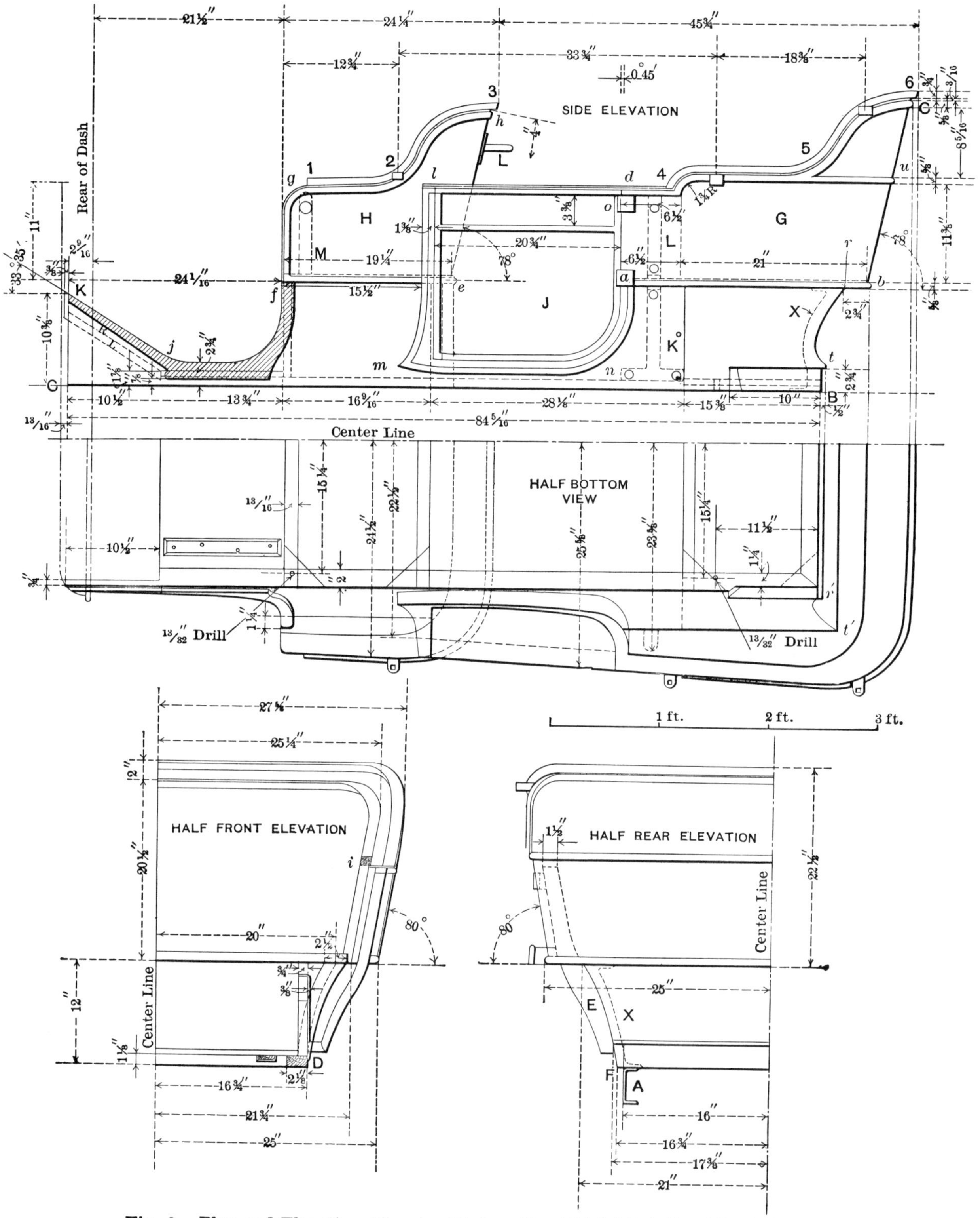

Fig. 2. Plan and Elevations Showing Various Sheet Metal Details of the Automobile

G shows the side elevation of the rear seat and H the side elevation of the front seat, both seats having an angle of 78° at the rear as shown. To secure rigidity, the seats have a bead along the bottom, *a b* on the rear and *f e* on the front seat. A similar bead is placed on the rear and front seats along the top as shown by C *d u* and *g h* respectively. The figures 1, 2, 3 and 4, 5, 6, in the two seats show an ash frame, which is secured to the top edge of the metal seats, the section of this ash frame being shown by *i* in the half front elevation, The shaded portion *f j* K in the side elevation shows the solid cast aluminum trimming, which is riveted to the sheet metal work. By J is indicated the door with special T-shaped trimmings around *l m n o*. Wrought iron braces K° L and M are secured to the inner side of the sheet metal body and seats to obtain rigid construction. Special cast malleable iron angles are placed in the corners of the body as shown by XX in the sides and rear elevations, which are riveted to the sheet metal work to secure stiffness.

After the scale drawing is received, the various full-size details are then laid out, and from these details the patterns are developed. The body being the most conspicuous feature of the automobile, great care should be exercised in laying out the detail and developing the patterns, and as the bodies are built on combination wood and steel skeletons, rivet, nail or bolt holes should be punched where possible before the patterns are formed up.

PATTERNS FOR TYPICAL AUTOMOBILE BODY

In Fig. 3 is shown one of the various styles of automobile bodies, with provisions for receiving a front seat on G F and a rear seat on D 1°. The side elevation is shown by A 6° 1° D E F G H *b*°, and the one-half rear elevation by 1 6 K J. In this case the lower rear part in the side elevation at 5^v 5^a 6^a 6° is recessed to clear the rear springs of the automobile, as deep as indicated by 5 5^x 6 in the half rear elevation.

The details of the side and rear having been properly drawn, the first step is to space the various profiles properly as follows: Divide the profile in the side elevation into a number of spaces, as shown by 1° 2° 3° and 4°, and number the corners of the profile 5°, 5^v and 6°, as shown. Through these small figures from 1° to 6° draw lines to the right, cutting the profile in the rear elevation at 1 2 3 4 5 5^x and 6. Establish an extra point in the deep curve between 3° and 4° in the side elevation as *a*°, which project to the right, thus obtaining *a* in the rear elevation. From the various intersections 1 to 6 in the rear elevation, draw lines indefinitely throughout the side, cutting the openings in the body at D E F and G H b°. From b° project a line to the right, cutting the profile in the rear elevation at *b*.

The patterns are now in order, and the one for the side will be laid out first. Take the girth of the profile in the rear elevation, from 1 to 6, measuring each space separately, as they are all unequal, and place them on the vertical line L M,

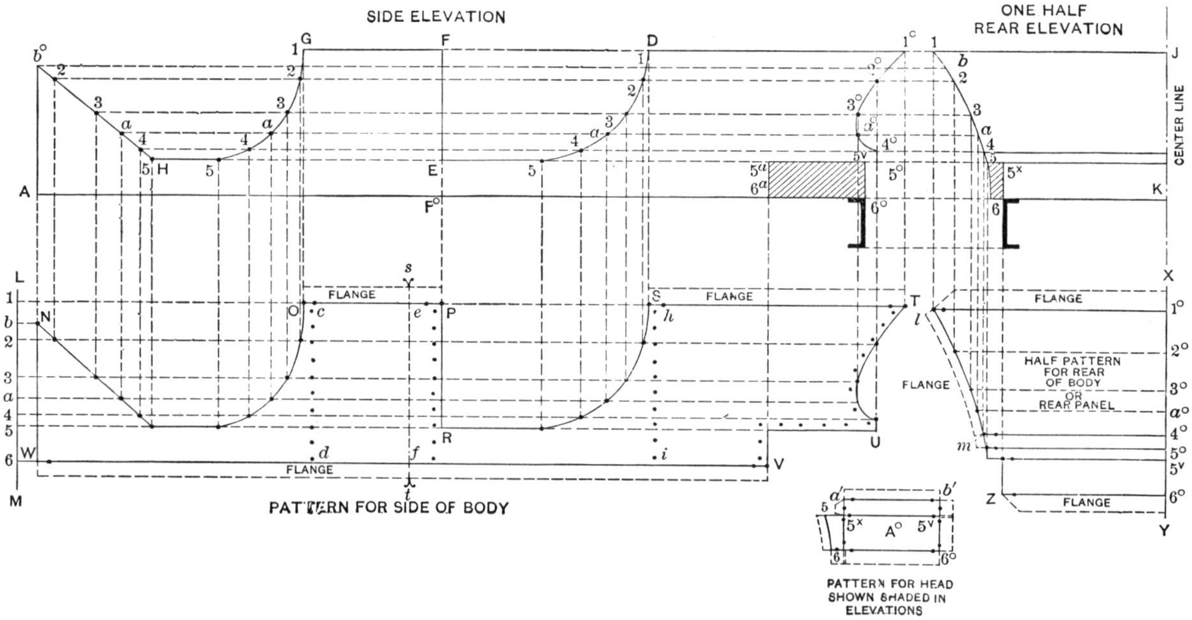

Fig. 3. Side and Rear Elevation of Automobile Body with Necessary Patterns

as shown by similar numbers and letters. Through these points, at right angles to L M, draw lines, which intersect by lines drawn parallel to L M from similar numbered intersections along the outline in the side elevation. A line traced through points thus obtained, as shown by N O P R S T U V W will be the pattern for the side of the body.

Flanges are allowed as indicated by the dotted lines. The flange along the bottom is usually made 1½ in. wide, turned toward the inside and nailed to the ash frame, then finally bolted to the channel frame of the chassis. The upper flanges, O P and S T, are also made 1½ in. wide, turned toward the inside, and reinforced with angle plates, previous to receiving the seats. The small dots shown in the side pattern indicate the rivet holes punched in the metal before forming up. Those along *c d* are used for riveting the front heel panel in place; those along *e f* for the front panel; those along *h i* for the rear heel panel, and those along T U V for the rear panel.

To obtain the pattern for the rear panel take the girth of the various spaces contained in the profile from 1° to 6° in the side elevation, and place them as shown by similar numbers and letters on the vertical line X Y. Through the small figures at right angles to X Y draw lines as shown, which intersect by lines drawn parallel to X Y from similar intersections in the profile 1 6 in the rear elevation. Trace a line through points thus obtained, then X T Z Y will be the one-half pattern, which is reversed opposite the line X Y for the full pattern. Flanges are allowed at top and bottom, as shown; also a 1-in. flange along *l m* as shown. This flange *l m* is not notched when riveting along T U in the side pattern, but is stretched to the required shape in a manner that is to be described.

Fig. 4. View in Corner of Sheet Metal Department

The pattern for the head shown shaded in the two elevations is shown by A° and is obtained as follows: Take a tracing of 5 5^x 6 in the rear elevation and place it as shown by 5 5^x 6 in A°. Complete the rectangle 5^x 5^v 6°, similar to

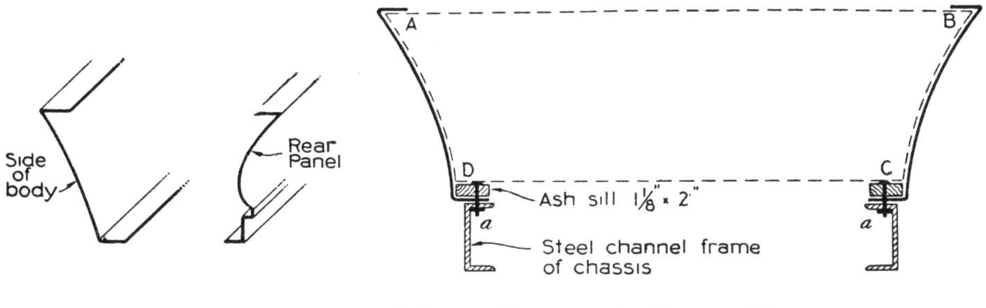

Fig. 5. Details of Curves Necessary in Sheets.—Fig. 6

5^a 5^v 6° 6^a in the side elevation. Now take the distance 5^x 5 in A° and place it as shown by 5^x a' and 5^v b'. Draw a line from a' to b' and add laps, as shown by the dotted lines.

In the large automobile works various models are manufactured requiring a different pattern for each model. When the patterns have been accurately laid out they are numbered and marked and can be saved for future orders. In the patterns just laid out no flanges are allowed along W N, N O or P R S. In this model the edge W N will be fastened to the dash, while N O and P R S will be re-enforced by means of solid cast aluminum mouldings, fastened in a manner to be described.

The patterns are laid out on No. 26 sheet iron, but in making up the body No. 20 sheet steel is used or pure sheet aluminum or aluminum coated sheets. Whatever metal is used it is usually obtained in car load lots direct from the mill to the shop. Now knowing the width of the sheet required from *s* to *t*, stock is ordered from the mill so as to have as little waste as possible; as a rule the sheets must all be trimmed, so that use is made of the power squaring shears, whose gauge is set equal to *s t* and all sheets cut this width, after which the pattern is laid over each sheet and marked.

Cutting and Punching the Body

The parts shown by N O, P R S and T U V are cut with the hand shears, and the rivet holes are punched by a special machine made for that purpose. When these bodies are cut, hundreds are cut at one time, and to facilitate matters the sheets are taken to the shears on a hand-car. When they are all cut they are again placed on the hand-car and taken to the punch machine, making use of the

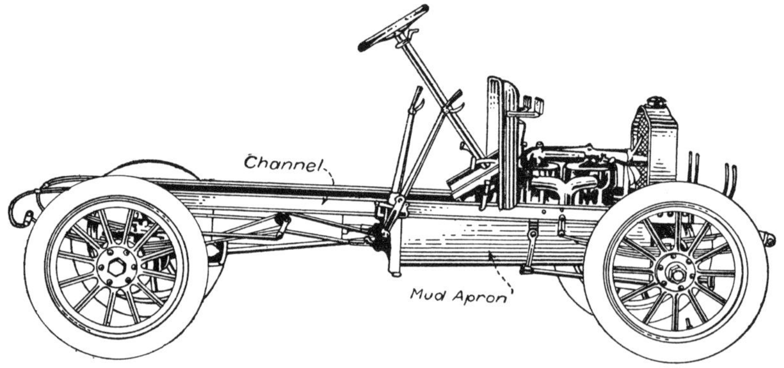

Fig. 7. General View of the Chassis of an Automobile

track laid in the concrete floor in and around the shop to the various machines, as shown in Fig. 4, which gives a part view of an automobile plant showing a number of completed bodies minus the seats, placed over one another with some seats, in the foreground, all made of sheet metal.

From the punch machine the sheets are taken to the power rollers and passed through a number of times, so as to smooth the metal, take out all buckles and make the metal more pliable. The sheets are now formed to their required curve on the large power rollers, after which they are carted to the large power brake and the necessary bends made, as indicated in Fig. 5, which shows the formation of the side and rear of the body. Before explaining how the metal body is joined together, the pattern for the front panel, also the front and rear heel panels, will hereafter be described.

Fastening Body to Chassis

As an allowance must be made on the patterns for these panels for the thickness of the ash sill, which is placed on the lower flange of the body, Fig. 6 has been prepared so as to make this clear. It shows a vertical section through F F° in Fig. 3, showing the method of fastening the metal body to the ash sill and channel of the chassis. In practice after the metal body has been put together, the lower flange of the metal body is nailed to the ash sill with 1-in. No. 14 flat-head wire nails, after which it is fitted to the channel frame of the chassis, but is not bolted to the chassis shown in Fig. 7 until the entire sheet metal work has been enameled in the enameling room. In fact, all the work above the chassis is finished complete in color and upholstering before it is bolted to the chassis, and as the bodies are usually numbered no mistake can occur.

PATTERNS FOR FRONT AND REAR PANELS

By making an allowance for the ash sill in Fig. 6, the patterns for the panels would be simply a reproduction of A B C D, which is shown by similar letters in Fig. 8, to which a 1½-in. flange has been added to the top and the bottom and a 1-in. flange on the curved part meeting the side of the body. A ½-in. beaded panel is to run through the face as shown, the beads being raised. This beaded finish is used for the front panel.

The pattern for the front and rear heel panels is shown by similar letters in Fig. 9, but contains a different design of panel, the arc being struck from a center, which is obtained as follows: Establish *a* on the center line as shown, also locate the point *b*. Now using *b* as a center and *b a* as

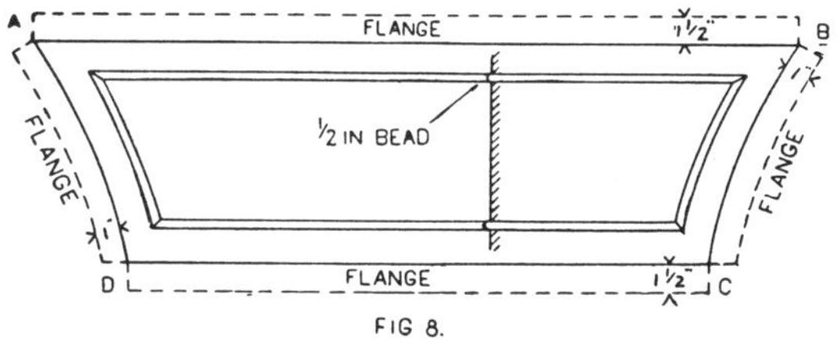

FIG 8.

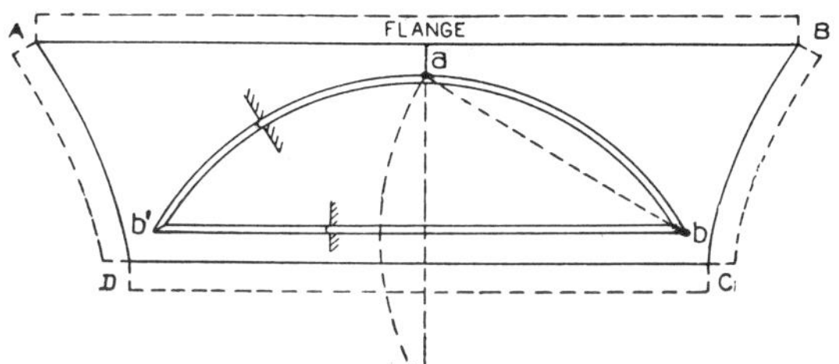

FIG 9

Patterns for Front and Rear Panels

radius, describe an arc, meeting a vertical line drawn from *a* at *c*. Now using *c* as a center draw the arc *b a b'*, which forms a guide line when beading. Flanges are allowed as shown by the dotted lines.

Constructing the Panels and Body

Where a large power drop press is used, these panels, as well as the flanges, are pressed in one piece, but where a power press is not used the beaded panels are made on a No. 1 power swedger or beader, using a ½ or ⅝-in. bead. When using the beader, the corners or miters of the beaded panel cannot be finished complete, but can only be beaded as far, in the front panel, as shown by *a* and *b* in Fig. 10. These corners are completed by using a set of steel die plates as shown in Fig. 11. These are usually made by the machinist in the automobile shop to suit the design desired, and a swedging hammer is made special for this purpose, as shown in Fig. 12, in which the bead *a* runs parallel to the handle, while the bead *b* runs at right angles to the handle.

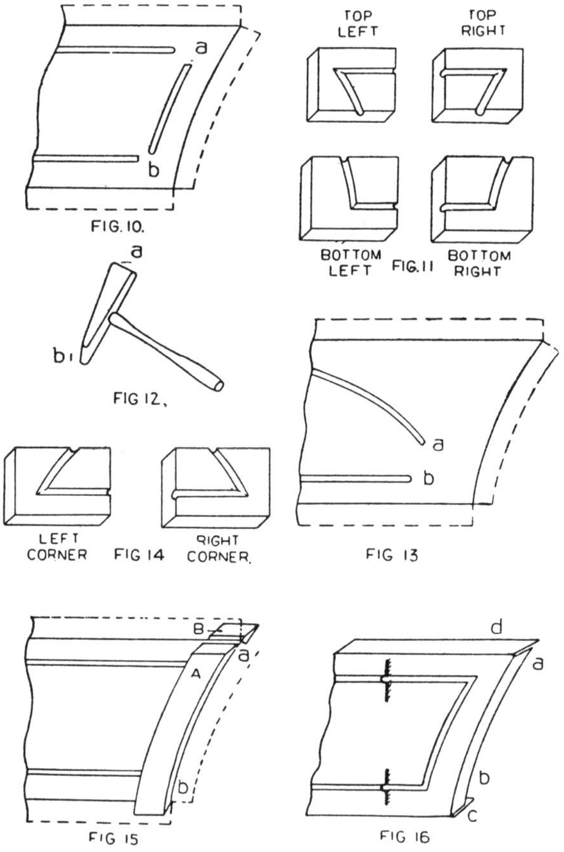

Details of Construction of Panels

When finishing these corners, the face *a* in Fig. 10 would be laid on the top left die in Fig. 11, and using the hammer, in Fig. 12, the bead is carefully worked out. In the heel panels the bead could only be beaded as far as *a* and *b* in Fig. 13, and then the beads are mitered by using the right and left die plates shown in Fig. 14, again using the hammer shown in Fig. 12, to make a finish.

When the curved flanges of the panels are to be turned over by hand, as it is usually done, this is accomplished by using a pair of steel stakes, which are made to suit the proper curvature, and are shown in position by A and B in Fig. 15. These two stakes are set in position, with the metal between them and the stakes clamped or fastened in a vise, after which the flange *a b* is turned over with a flanging hammer, so that when complete it will look as shown by *a b* in Fig. 16.

The lower flange *c* and upper flange *d* are bent on the brake. In a similar manner, using a different shaped stake of the required curve, is the flange *a b* in the rear panel in Fig. 17 turned outward. As before mentioned, none of these stakes are necessary when the various metal parts are pressed on the power drop press, which would make the flange without any hand work.

Fig. 17 also shows the constructive view, indicating how the body and the various panels are riveted together and the corners reinforced by gusset plates.

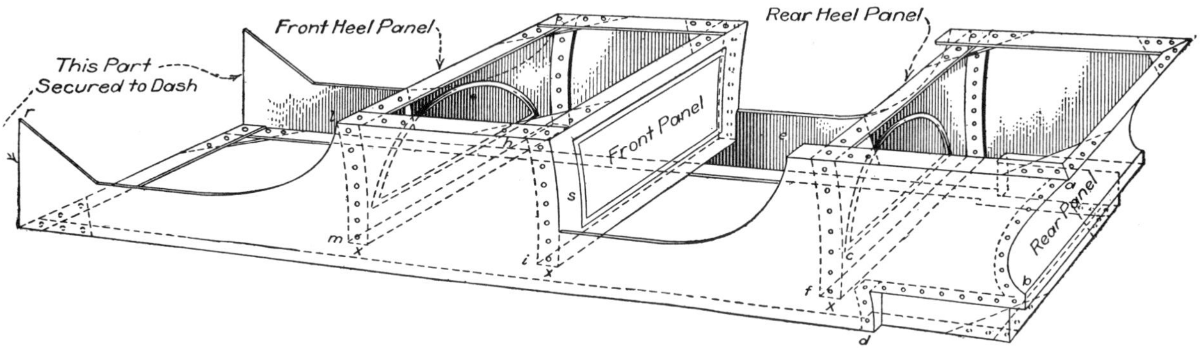

Fig. 17. Constructive View Showing How Body and Panels Are Riveted and Corners Reinforced

The rear panel is riveted along *a b c d*; the rear heel panel along *e f* ; the front panel along *h i*, and the front heel panel along *l m*. The corners are all reinforced by riveting flat gusset plates as indicated by *o o*, etc. In addition to the

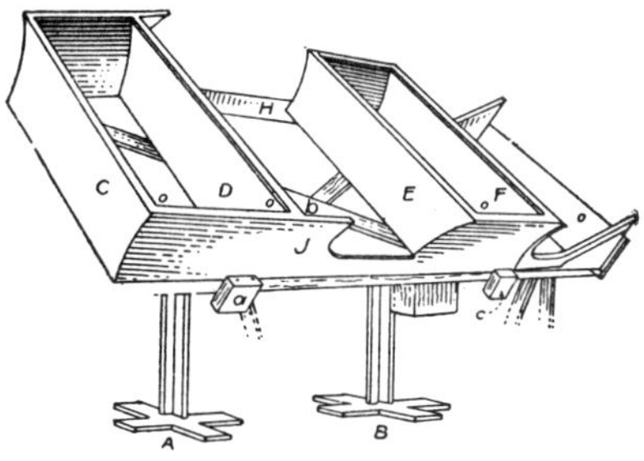

Fig. 18. Pivoted Frame Used in Building Automobile Bodies

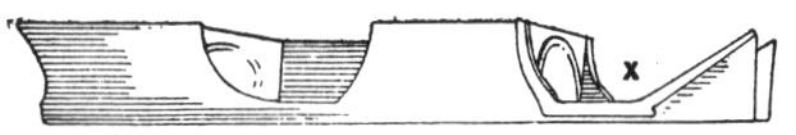

Fig. 19. Automobile Body in Construction Showing Heel Panels

gusset plates, iron angle plates are riveted in all corners as shown by X, X, L, K°, etc., in the scale drawing in Fig. 2, which makes the body perfectly rigid. In Fig. 17 X X X indicates the space allowed below the panels for the ash sill to pass under.

When building up these metal bodies, a wooden frame is used, having the exact dimensions of the channel of the chassis on which the body is to be placed and is shown in Fig. 18. This wooden frame *a b c* rests on and is pivoted on two cast iron stands A and B, which are secured to the floor, so that when riveting the various panels C D E and F to the sides of the body H and J, the body can be turned up or down on either side to facilitate working. The picture shows the body slanting toward the reader. The corner gusset plates may be seen at *o o*, etc. Another view of the body in course of construction is given in Fig. 19, showing the front and rear heel panels as well as the cast aluminum molding X being fitted.

When riveting together the various parts of the body shown in Fig. 17, all the riveting is done from the outside, using 12-oz. tinned rivets and special shaped dollys and stakes. The rivet holes in the metal sides of the body are slightly countersunk, so that the rivets will show a smooth surface.

SOLDERING AND SMOOTHING THE RIVETED JOINTS

After all riveting is done the body is carted to a different part of the shop ready for soldering all rivet heads and joints. Before any soldering can be done to the black sheets the surface must be cleaned bright. This is quickly accom-

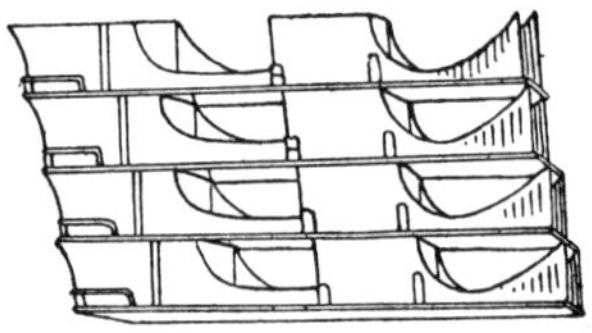

Fig. 20. Group of Bodies Showing Soldering Lines

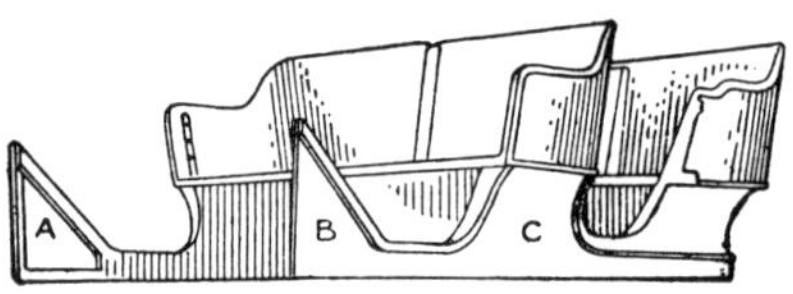

Fig. 21. Body with Cast Aluminum Moldings

plished by using an emery wheel with a handle attached and driven by electricity. This emery wheel contains a small motor with wires to attach to any electric light socket. In using, the handle is taken in hand, the emery wheel placed against the joint on the metal body, and by turning a switch alongside of the handle, power is obtained, and in a short time the joint is bright and smooth.

The seam is now brushed over with (Reade's) soldering flux and the joints thoroughly sweated with half-and-half solder, using 10-lb soldering coppers. This soldering flux has the preference because it will not rust the metal after drying. Fig. 20 shows a group of bodies just soldered, the white lines in the picture of the bodies showing the soldered joints.

The soldered joints are now cleaned and scraped smooth, after which the cast aluminum body moldings are fitted along the curves *e s* and *l r* in Fig. 17, as shown in the picture in Fig. 21 by A B C. These cast aluminum moldings are

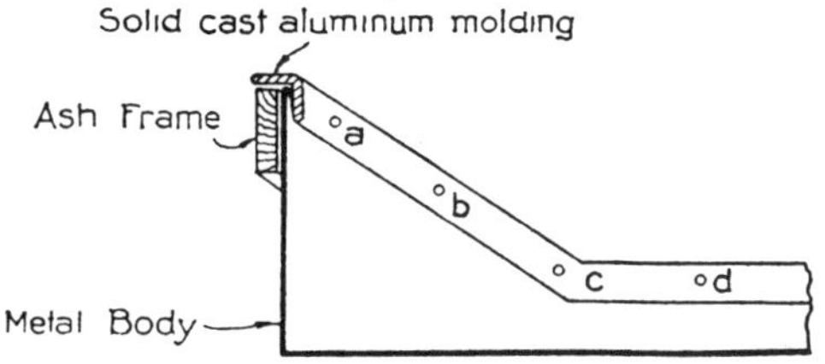

Fig. 22. Attachment of Cast Aluminum Molding

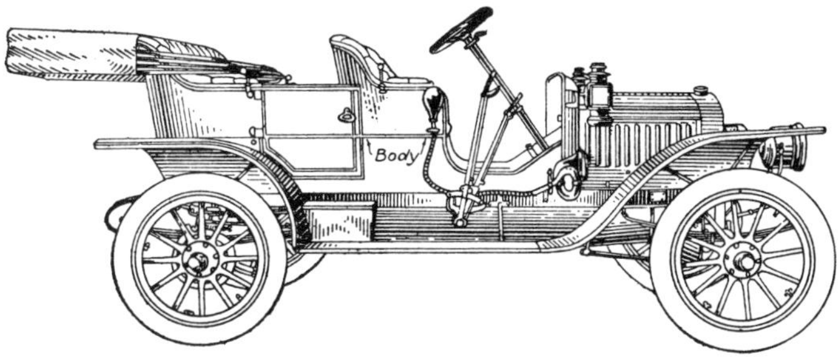

Fig. 23. One Type of Automobile to Which Patterns Relate

riveted to the raw edge of the steel body as shown in detail in Fig. 22 by *a b c d*, the molding having an angle formation, which rests on the raw edge of the steel body and is riveted smooth on the outside, after which the ash frame is set under the molding as shown, making an inside finish.

DEVELOPING PATTERNS FOR AUTOMOBILE BODIES WITH DIFFERENT PROFILES

The patterns and description of the body just given are used on automobiles of the type shown in Fig. 23. Fig. 24 shows the principles of parallel line development applied to a body the profile of which has a double curve. In this case 1 5 B A shows the half rear elevation and C D E F G H J K L the side elevation, the rear of the car being shown by C D and the line of the dash by E F. As the profile of the rear C D is similar to the profile of the side 1 5, then the miter cut for the one will answer for the other. The first step is to project F and G in the side elevation to the profile in the half rear elevation and obtain points 2 and 4 respectively. Establish an intermediate point as 3, which project into the side elevation as shown. As the curve in G H is very deep, establish another point as *a*, which project to the half rear elevation, thus obtaining *a′*. The number of intermediate spaces required in the profile 1 5 all depends upon the size of the curves.

Having spaced the profile 1 5 as desired, the patterns are laid out in the usual way, the girth of the profile 1 5 being placed on vertical line M N chosen at will.

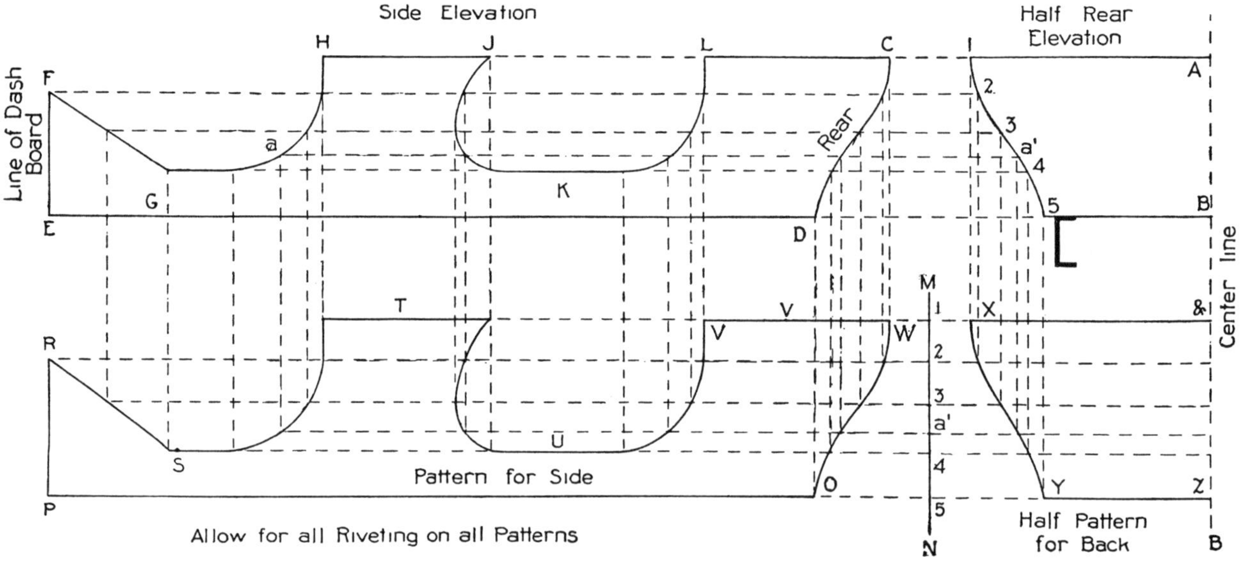

Fig. 24. Development of Body of the Automobile

Riveting flanges must be allowed on all patterns, as before explained. No matter what profile the body may have, whether a pitch or a convex or concave curve or a double curve, the parallel line method can be used in developing the pattern.

VARIOUS TYPES OF RADIATORS

Some bodies of automobiles are so constructed that the hood over the dash board is connected to the side of the body in one piece and is known as the dash hood, or sometimes called the dash shield. This construction usually takes place in touring cars, raceabouts, roadsters, semi-racers, tourabouts, surreys, racing cars, also those of the torpedo and gunboat types. The methods of laying out these bodies with dash hoods attached will be taken up in regular order, 16 types of cars having been selected. These studies will give the rules for laying out any shape body which may arise in automobile construction. Some bodies will have straight sides, as well as curved, and the principles will be given for developing the dash hoods by parallel lines and radial lines, as well as by triangulation. Where possible the hoods should be laid out by the simple rule, which is by parallel lines, but where the shapes are such that parallel lines cannot be used, then radial lines or triangulation must be used.

The lower end of the dash hood is always obtained from the dashboard, while the shape of the dashboard is usually obtained from the shape of the radiator. Thus the radiator forms the basis for the outline of the dashboard and lower end of the hood. Three of the many-shaped radiators are shown in Fig. 25 by X, Y and Z.

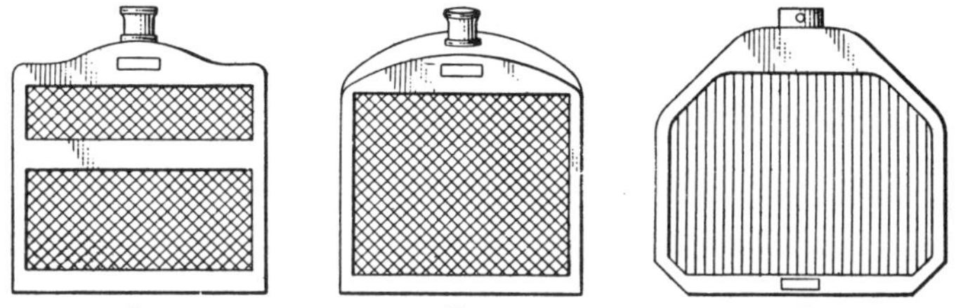

Fig. 25. Three Shapes of Radiators

PATTERNS FOR BODY AND DASH HOOD COMBINED IN AN AUTOMOBILE WHOSE SIDES ARE STRAIGHT, THE HOOD BEING ON A HORIZONTAL PLANE

In Fig. 27 is shown a view of an automobile on which the body and dash hood are combined. Since the hood is on a horizontal plane, the pattern can be laid out by means of parallel lines the same as in pipe work. In this case the sides of the body are straight and need no development. These sides and dash hood are so laid out that a seam takes place at the top at A.

The method of development is shown in Fig. 28, in which the side elevation of the body and dash hood is shown, as well as the front elevation of the radiator and dashboard and hood. The development of a pattern of this kind is very simple. Divide the curved part of the outline of the dashboard into equal spaces, as shown by the small figures 1 to 4, from which points, parallel to A 5°, draw lines until they intersect the curve of the dash hood *a* F E, as shown. As the side of the body is straight and to avoid a confusion of lines, take a tracing of A B C D E F 4° 5° and place it as shown to insure by similar letters and figures below the elevation from A^1 to 5′. Starting from 4′ set off the girth of the dashboard; at right angles to 1′ 4′ and through these figures draw lines, which intersect by lines drawn, parallel to 1′ 4′ from similar intersections on the curve *a* F in the side elevation. A line through the points obtained will be the half pattern of the dash hood, combined to the body on the line F^1 4′ as shown. Riveting flanges must be added as required and as explained.

The pattern for the rear panel A B in the side elevation is simply the girth of A B and in width equal to 5 *b* in the front elevation. The raw edge of the dash hood along E^1 F^1 a' in the pattern or along A B in Fig. 27 is usually finished with a hollow tubing, round or angular bead or angle, as shown in the four cuts in Fig. 29. The first cut shows the brass tubing A with a slot cut in it to admit the sheet metal of the dash hood B, the tubing being soldered at *a*. As a rule brass tubing is used ½ in. inside diameter with 1-16 in. wall, making the tubing ⅝ in. outside diameter.

Bending and Slotting the Tubing for Dash Hoods

The brass tubing is cut off to the required length or as long as twice the girth of E^1 F^1 a' in Fig. 28, allowing enough material at each end of the tube to allow it to be grasped and trimmed and filed. The one end of the tube is closed and the

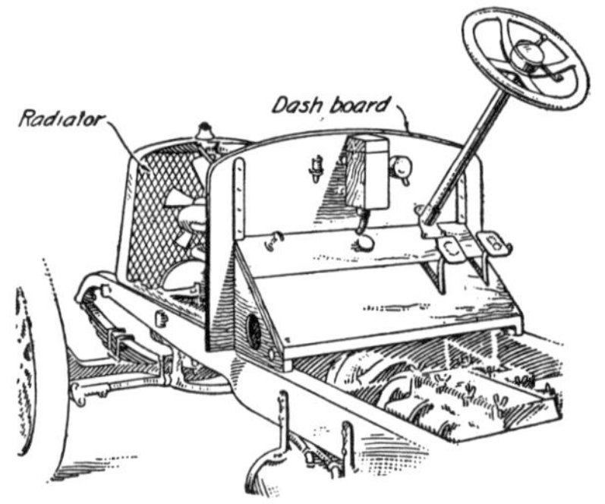

Fig. 26. Radiator in Position

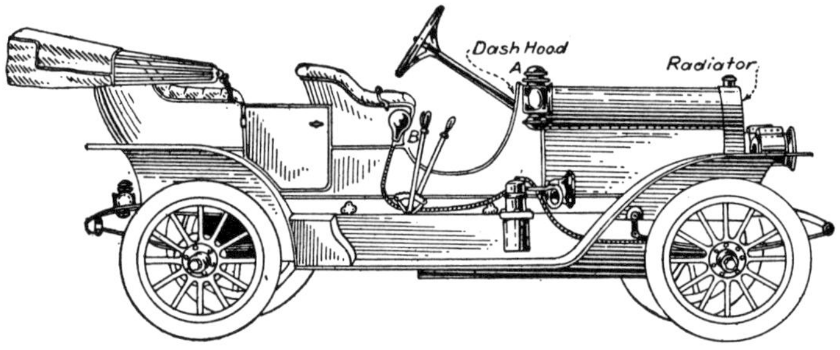

Fig. 27. View of Automobile Body with Dash Hood Attached

tube filled with molten block tin, and when cooled the slot shown by *a* in Fig. 29 is cut by using a No. 0 plain milling machine, shown in Fig. 30, in which A shows the table with an automatic feed, which can be raised or lowered, and B is a metal saw used for cutting the slot. The tubing is now bent to conform to A B, Fig. 27, care being taken when bending that the slot will fit over the raw edge of the sheet metal dash hood.

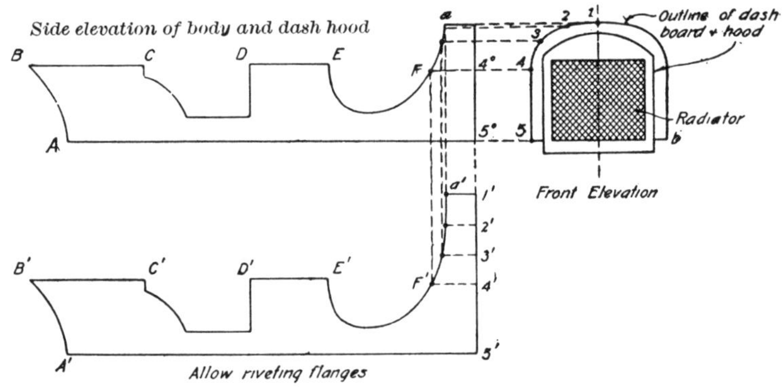

Fig. 28. Pattern for Body and Dash Hood

After the tube has been properly bent, the block tin is melted out of the tubing by heating it with a blow torch.

When soldering this tubing to the metal sheets, the surface of the metal of the dash hood B in Fig. 29 is properly tinned, as described, when the brass tubing is

Fig. 29. Finishing Raw Edge of Dash Hood

sweated with solder, using the flux previously mentioned. The joints are scraped smooth previous to enameling and if the brass tubing is to be kept polished, solder on it must be carefully scraped and the brass work cleaned and polished. Black soft steel tubing can also be used in a similar manner.

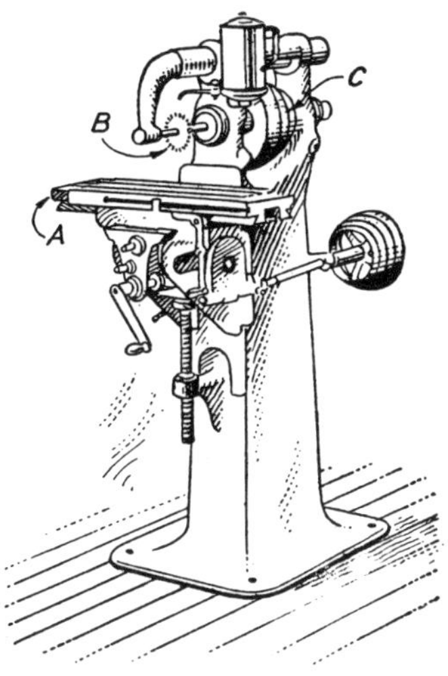

Fig. 30. Milling Machine

The second method of finishing the raw edge of the dash hood is to use either cast aluminum or soft band iron of the circular section, shown by C. This band is riveted at intervals, as at *a*, to the metal dash hood D, countersinking the holes so as to rivet smooth on the outside, the rivets being filed off to make a clean, smooth job. In similar manner in the third method is the angular finish E riveted to F at *a*. Care must be taken in bending these bands C and E to have the true shape of the finished dash hood. In the fourth method an angle shape L is used, as shown by G, riveting to the sheet metal H, at *a*.

PATTERN FOR BODY AND DASH HOOD COMBINED FOR TOURING CAR, THE SIDES OF THE BODY BEING STRAIGHT AND THE DASH HOOD BEING DEVELOPED BY THE PARALLEL LINE METHOD

In Fig. 31 is shown a view of a touring car, in which the body and hood combines. The dash hood is angular in shape, runs at an angle to the horizontal and can be developed by means of parallel lines, but must be raked, which means that a true profile of the dash hood must be found at right angles to the pitch of the dash hood.

This is made clear in Fig. 32, in which A D B C 4° 2° 2^v shows the side elevation of the body and dash hood. The radiator and the outline of the dashboard and hood are shown in the front elevation. When drawing the front and side elevations, care must be taken to have the outline of the dash hood 1, 2, 3, 4, in line with the dashboard 1° 4°, in the side elevation, as shown. The pitch and depth of the dash hood, 1° 1^v, may be established at pleasure, and then a graceful curve may be drawn from 1^v to a^o.

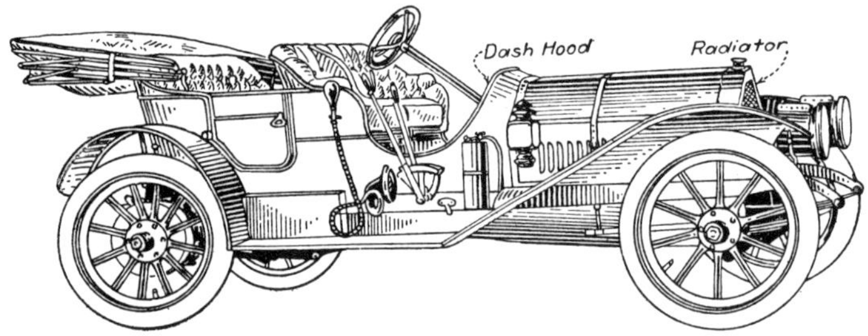

Fig. 31. Touring Car with Dash Hood and Body combined

As the outline of the dash hood represents the true section on the vertical plane 1°-4°, in the side elevation; then, before the pattern for the dash hood can be obtained, the true profile at right angles to the pitch of the hood, or on the line *c d*, must first be obtained as follows: From the various intersections, 1, 2 and 3, in the outline of the dash hood, draw horizontal lines, until they intersect the vertical line in the side elevation at 1° 2° and 3°, from which points, parallel to the pitch, 1° 1^v, draw lines indefinitely, as shown, intersecting the upper line of the hood at 1^v, 2^v and 3^v.

At right angles to the lines just drawn, draw any line, as *a′ b′*. Measuring from the line *a b*, in the front elevation, take the various distances to points 2 and 3 and place them on similar lines in the true profile, measuring, in each instance, on either side of the line *a′ b′*. Draw lines through points thus obtained; then 3′, 1′, 3′ will be the true profile of the dash hood on the line *c d* in elevation.

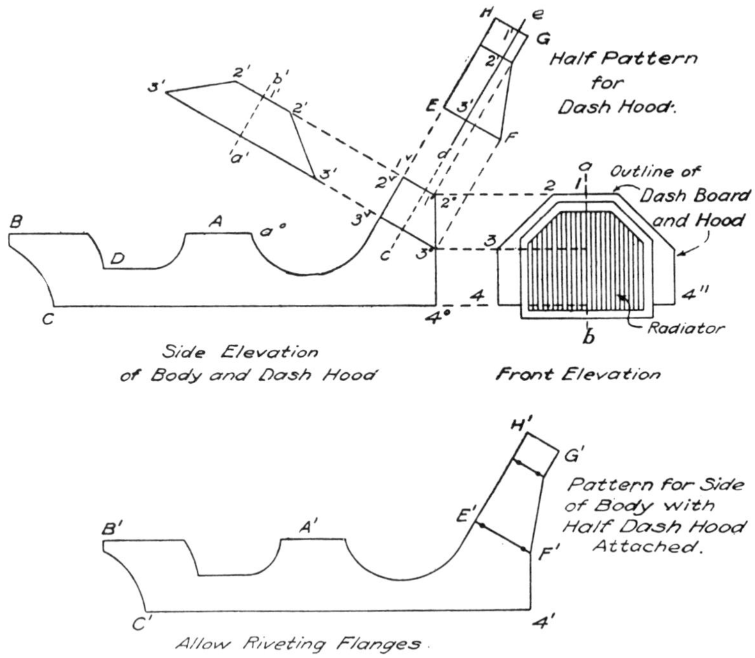

Fig. 32. Pattern for Body and Dash Hood shown in Fig. 31

The pattern for the dash hood is now in order, and is obtained by first drawing any line, as *d e*, at right angles to the pitch of the hood. On this is placed the half girth of the true profile, as shown from 1′ to 2′ to 3′ on the line *e d*. At right angles to *d e*, through the small figures on it, draw lines, which intersect by lines drawn parallel to *d e* from similar numbered intersections on 1° 4° and 1^v 3^v. A line traced through points thus obtained, as shown by E F G H, will be the half pattern for the dash hood. As the body has straight sides, take a reproduction of 3^v A B C 4° 3° in the side elevation and place it as shown by E^1 A^1 B^1 C^1 4′ F^1 in the pattern, and on the line E^1 F^1 place a reproduction of E F G H, or the half dash hood pattern, as shown by E^1 F^1 G^1 H^1. Then G^1 H^1 B^1 C^1 4′ will be the pattern for the side of the body with the half dash hood attached, which gives a seam along 1° 1^v in the side elevation.

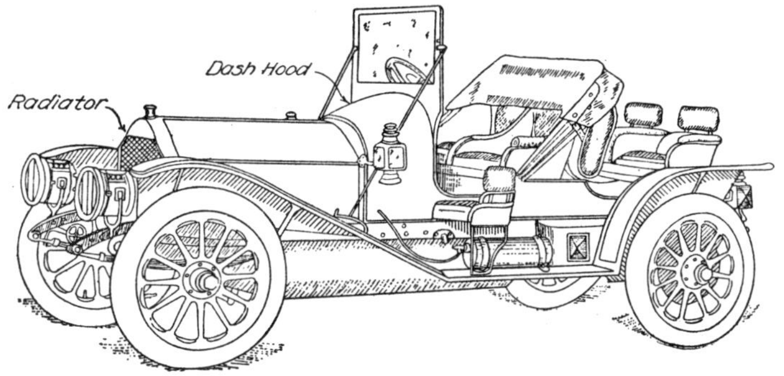

Fig. 33. Rumble Raceabout, with Dash Hood and Body Combined

The pattern for the rear panel B C is obtained by using the girth, B C, and the width, 4 4″, in the front elevation. Riveting flanges must be allowed, making a finish along the upper edge of the dash hood with brass or aluminum moldings, as before explained.

PATTERN FOR BODY AND DASH HOOD COMBINED FOR A RACEABOUT, THE SIDES OF THE BODY BEING STRAIGHT, THE DASH HOOD BEING DEVELOPED BY PARALLEL LINES

Fig. 33 shows another type of automobile, known as a rumble raceabout, on which, as in the previous cars, the dash hood is joined to the body in one, except in this case the dash hood is cut off on two vertical planes. The pattern for this dash hood or shield can be developed by parallel lines, but requires raking, or finding the true profile at right angles to the rake or pitch of the hood.

The method of finding this body pattern and dash hood is shown in Fig. 34, in which A, B, C, D, E, F, 1, 1 shows the side elevation of the body and dash hood, the radiator and outline of dash board and hood being shown in the front elevation, which is placed in line with the side view.

The first step is to divide the curved part of the half outline of the dash hood into equal spaces, as shown by the small figures, 1 to 4, in the front elevation, through which points horizontal lines are drawn, cutting the center line *a b* in the front elevation, also the line of the dash board 1 A in the side elevation, as shown by the small figures 1 to 4. From these points, 1 to 4, on the line 1 A, draw lines, indefinitely, parallel to 1, 1, cutting the upper plane of the dash hood, as shown by similar numbers.

At pleasure, draw any line, as *a′ b′* perpendicular to the lines just drawn, and, measuring from the line *a b*, in the front elevation, take the various distances to points 1, 2, 3 and 4 and place them on each side of the line *a′ b′* in the true

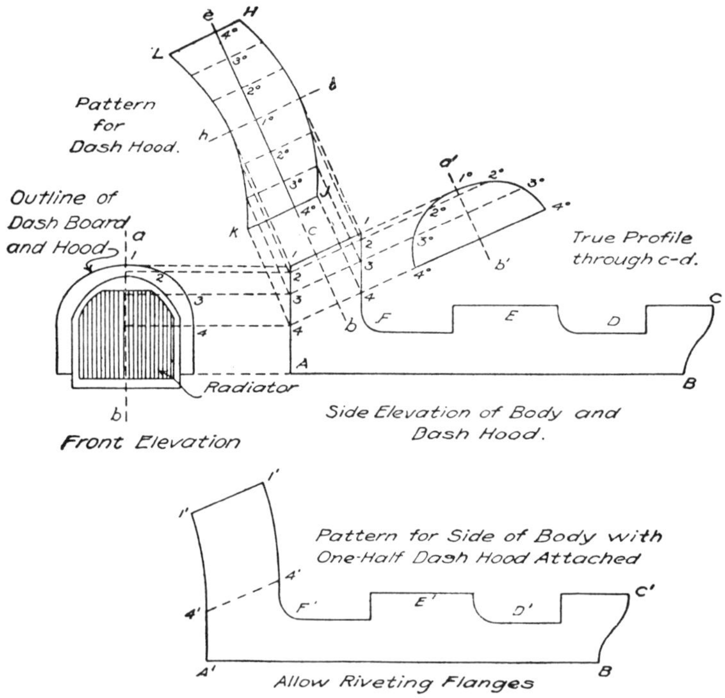

Fig. 34. Pattern for Dash Hood and Body Shown in Fig. 33

profile on similar numbered lines, thus obtaining the points of intersections 1° to 4° on both sides. A line traced through these intersections will give the true profile, through *c d*, which is at right angles to 1, 1.

Extend *b c* as *c e*, and upon it place the girth of the true profile, as shown by similar numbers. Through these figures, at right angles to *e c*, draw lines, which intersect by lines drawn parallel to *e c* from similar numbered intersections on the two vertical planes 1, 4. Trace a line through points thus obtained, then J K L H will be the full pattern for the dash hood.

As the sides of the body are straight, take a tracing of 4 A B C D E F 4 and reproduce it as shown by 4′ A^1 B^1 C^1 D^1 E^1 F^1 4′ in the pattern and on the line 4′ 4′ place a reproduction of the half pattern, J K *h i*, as shown by 4′ 4′ 1′ 1′ in the pattern. Then 1′ A^1 B^1 C^1 4′ 1′ is the pattern for side of body with one half dash hood attached. Nailing and riveting flanges must be allowed. The pattern for the rear panel, B C, in side elevation is obtained as previously described.

PATTERN FOR BODY AND DASH HOOD OF A TOURING CAR WHOSE SIDES ARE STRAIGHT, THE OUTLINE OF THE DASH HOOD HAVING A DOUBLE CURVE AND BEING DEVELOPED BY PARALLEL LINES

Fig. 35 shows the usual type of touring car in which body and hood are combined. The outline of the dashboard has a double curve, as shown in the front elevation in Fig. 36. It will be noticed that the hood in Fig. 35 has a bent corner. The pattern for the dashhood can be developed by parallel lines, but the true profile must be obtained by raking.

In Fig. 36 the side elevation of the body and hood is indicated by A B, 5, 1, 1, 5 C D. The sides of the body are straight. Place the front elevation in its proper position as shown, the outline of the dashboard being obtained from the outline of the radiator, as shown by the double curve 1 to 3 to 5. This half outline 1 to 5 is divided into equal parts. Through the points of division horizontal lines are drawn, intersecting the center line *a b* in the front elevation, also the line of the dashboard 1 B in the side elevation, as shown by the small figures 1 to 5, on 1 B.

Fig. 35. Touring Car with Combined Body and Hood

From these small figures 1 to 5 on 1 B draw lines parallel to the pitch of the hood 1-1, cutting the upper side of the hood also from 1 to 5 and extending the lines indefinitely as shown. At right angles to the lines just drawn, draw any straight line as *a′ b′*, which is used in obtaining the true profile of the dash hood. Measuring from the line *a b* in the front elevation take the various distances to points 2, 3, 4 and 5, and place them on lines drawn through similar numbers, measuring in each instance on each side of the line a' b', thus obtaining the intersections 5° to 1° to 5°. A line traced through these points will be the true profile at right angles to the pitch of the hood shown by c d in the side elevation.

For the pattern for the dash hood proceed as follows: Extend the line d c as c e, upon which place the girth of the true profile as shown from 5° to 1° to 5° on c e. Through these small figures at right angles to c e draw the usual measuring lines, which intersect by lines drawn parallel to e c from similar numbered intersections on the upper and lower lines of the dash hood in elevation. A line traced through points thus obtained as shown by E F G H will be the pattern for the dash hood.

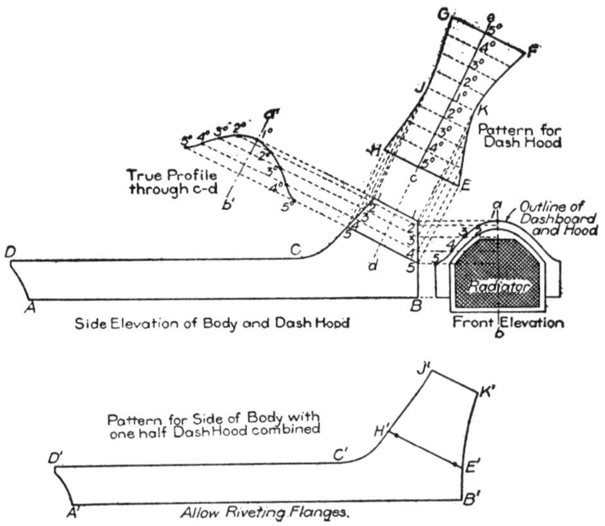

Fig. 36. Pattern of Body and Hood Shown in Fig. 35

As the sides of the body are vertical, a reproduction of A, B, 5, 5, C, D, is placed as shown by A^1, B^1, E^1, H^1, C^1, D^1, and on the line E^1, H^1 a tracing of the half pattern for hood H, E, K, J is placed as shown by H^1, E^1, K^1, J^1 in the pattern. Nailing and riveting flanges must be allowed as explained in the last pattern, finishing the raw edge of the dash hood along 1 C in side elevation with brass or aluminum trimmings as explained in a previous figure.

As but one-half of the dash hood pattern has been added to the body side, this brings the seam of the dash hood along 1, 1, in the side elevation. The rear panel D A is a rectangular piece of metal with flanges allowed. Its girth is equal to D A and the width equal to the full width of the dashboard in the front elevation.

PATTERN FOR BODY AND DASH HOOD OF A TORPEDO TYPE OF AUTOMOBILE HAVING STRAIGHT SIDES, THE HOOD BEING DEVELOPED BY PARALLEL LINES

In Fig. 37 is shown a view of the torpedo type of automobile, with straight sides and curved back, combined with dash hood, which can be developed by means of parallel lines, but requires raking or finding the true profile of the hood, at right angles to its pitch. As the rear corners of this body are round or contain the section of a quarter circle, allowance must be made in transferring the patterns for the body.

In Fig. 38 A, B, C, D, E, F, 5, 1, 1, 5, A shows the side elevation of the body and dash hood. The plan shows the rear corners of the body being indicated in diagram A^2, the quarter-round corner *i j* being shown, struck from the center point *h*. In its proper position to the left of the side elevation draw the front elevation, showing the shape of the radiator, from which the outline of the dashboard and hood is obtained. The dashboard or hood outline represents the true section on the dashboard line 1 A in the side elevation.

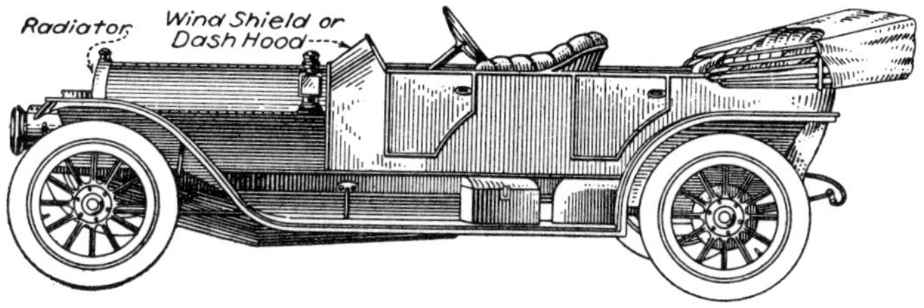

Fig. 37. Torpedo Type of Automobile

Divide the curved part of the dash hood in the front elevation into equal parts, as shown from 1 to 5, from which points horizontal lines are projected, cutting the dash lines, 1 A in side elevation from 1 to 5. From these intersections, parallel to the pitch of the hood 1-1, draw lines intersecting the upper line of the hood also from 1 to 5 and extend them indefinitely as shown. Perpendicular to the lines just drawn draw any lines as *a′ b′*. Now, measuring from the line *a b* in the front elevation, take the various projections to points 2, 3, 4 and 5 and place them on lines drawn from similar numbers in the side elevation, measuring on each side of the line *a′ b′*, and thus obtain the intersections 2°, 3°, 4° and 5°. The line through these points gives the true profile through *c d* in the side elevation or at right angles to the rake of the hood.

Extend *d c* as *d e* and on it place the girth of the true profile, measuring each space separately, as they are all unequal, and place them, as shown, from 5° to 1° to 5° on the line *c e*. Through these small figures, at right angles to *c e*, draw lines which intersect by lines drawn parallel to *c e* from similar numbered intersections on the upper and lower line of the dash hood in the side elevation. A line traced through points thus obtained will give the full pattern for the dash hood.

The sides of the torpedo body are straight, with rounded rear corners, as shown in the part plan A^2. In some cases the sides of the body are curved. Knowing the amount of the rounded corner *i j* in plan A^2, project the point *i* to the elevation at *l*. Set off this distance B^1 *l* as shown at the bottom by B *m* and take a tracing of B B^1 and place it from *m* to *l* as shown by the dotted line. In laying out the pattern for the side of the body, this portion *m l* B^1 B is omitted, for in its place the stamped rounded corner *i j* in plan is riveted.

Take a tracing of A m l C D E F 5, 5 and place it as shown by A^1 m' l' C^1 D^1 E^1 F^1 5′, 5′. On the line 5′ 5′ place a reproduction of the half dash hood pattern M N H L, placing H L on the line 5′ 5′ in the pattern, as shown by 5′ N^1 M^1 5′. Then A^1 m' l' N^1 M^1 will be the pattern for the body and hood combined.

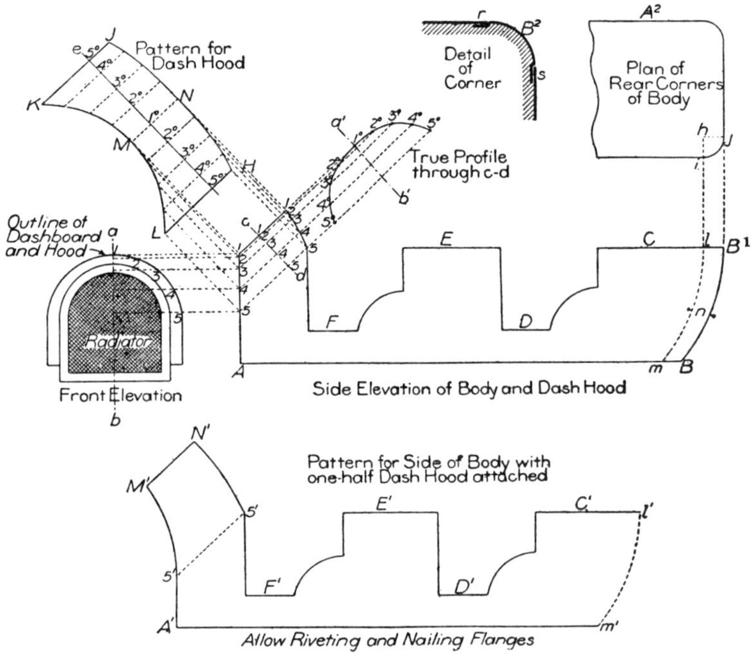

Fig. 38. Pattern for Body and Dash Hood Shown in Fig. 37

When laying out the pattern for the rear panel B^1 B in side elevation the amount n or the distance i h in plan must be deducted on each end of the rear panel to make allowance for inserting the rounded corner shown in detail by B^2. The rounded corner B^2 is stamped or pressed in a drop press with a counter-sunk edge as shown, so that when the side r and rear s is riveted in place it will show a smooth, clean joint. As previously explained, the metal is first tinned, then the corners riveted and then soaked with solder and filed smooth before enameling.

PATTERN FOR BODY AND DASH HOOD OF A TOURABOUT WITH STRAIGHT SIDES, THE DASH HOOD REQUIRING TRIANGULATION IN ITS SOLUTION

In Fig. 39 is shown a view of a tourabout with straight sides combined with dash hood. The shape on each end of the hood is such that the method of triangulation must be employed in solving the pattern. The peculiar shapes are shown in the illustration in Fig. 40, in which A B 1 8 C D E shows the side elevation of the straight side of the body and 1 5 6 8 the side elevation of the attached dash hood.

In its proper position, in line with the side elevation, draw the shape radiator wanted. From this are obtained the outline of the dash board as well as the lower part of hood on the line 5 B in side elevation. From 6 in the side elevation draw a line until it intersects the center line *a b* in the front elevation also at 6, and from this point complete at pleasure the outline of the upper part of the hood shown from 6 to 7 to 8 to 1 which shows a vertical section on the line 6-8-*s*.

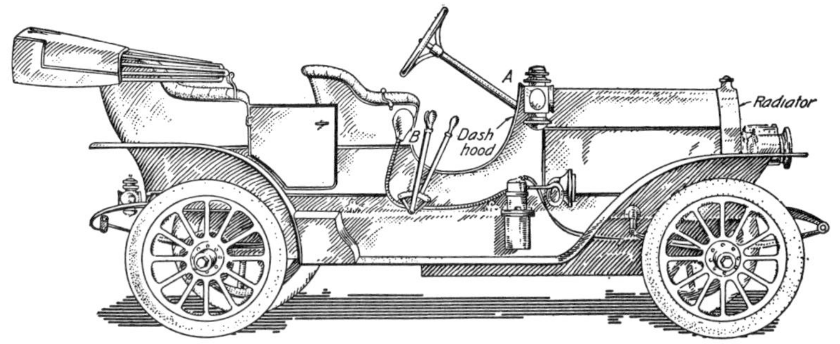

Fig. 39. View of a Tourabout

This dash hood is developed as any transition piece in sheet metal work, the section of which, on the line 5 1 in the side elevation, is shown by 5 1 in the front elevation, and where the vertical section on the curved line 6 8 *s* in the side elevation, is shown by the irregular line 6 7 8 1 in the front elevation. It should be understood that in this problem, as well as in others which follow, the two sections shown in the front elevation can be drawn at pleasure. The first or lower section 5 1 being drawn to conform to the shape of the desired radiator and the upper section 6 7 8 1 is drawn to suit the pitch of the hood 5 6 in the side elevation, this vertical hight being indicated by 5 6 in the front elevation. The shape 6 8 *s* in the side elevation can also be drawn at will. Sometimes the hood has a taper, starting from the line 5 B in the side elevation and getting wider as it approaches to the widest part of the body at *s*.

The first step in developing the dash hood under the conditions shown is to divide the lower outline in the front elevation into equal spaces, as shown from 1 to 5, from which points project lines to the proper line in the side elevation as shown also from 1 to 5. Number the corners in the upper outline in the front elevation 6 7 and 8 and also project these lines into the side elevation as shown by 6 7 and 8. As 6-7, 7-8 and 8-1 in the front elevation represent flat surfaces, draw lines from 1, 2 and 3 to 8 and from 3, 4 and 5 to 7. These lines then represent the bases of triangles which will be constructed; the altitudes or heights will be equal to the horizontal distances between similarly numbered points in the side elevation.

To obtain these triangles, which will show the true lengths, proceed as follows: Extend the line B 5 in the side elevation as shown by B *d*, and starting from any point as 7°, set off the distances of 7-5, 7-4 and 7-3, in front elevation as shown by 7°-5, 7°-4 and 7°-3 on the line B *d*. From 7° draw a perpendicular line to B *d* and intersect it by a vertical line erected from 7, thus obtaining 7′.

Draw lines from 4, 5 and 3 to 7′. These show the true lengths of the lines having similar numbers in the front or side elevation. In a similar manner take the lengths of 8-2 and 8-3 in the front elevation and place on the line B *d* from 8° to 2 and 8° to 3. From 8 erect a vertical line, meeting the horizontal line drawn from 8° at 8′. Draw lines from 2 and 3 to 8′ which are the true lengths of similarly numbered lines in elevations.

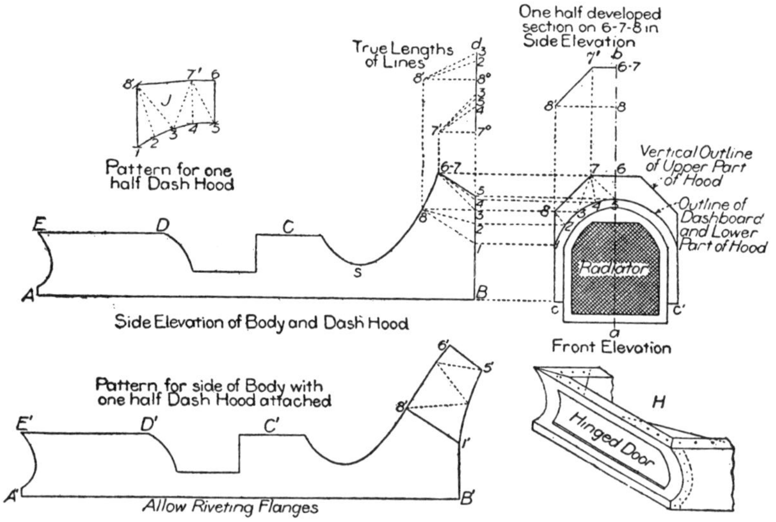

Fig. 40. Pattern for Body and Dash Hood, Shown in Fig. 39

As 6-7-8 in the front elevation shows but a vertical section on the line 6-7-8 in the side view, a true section on 6, 7, 8 in the side elevation must be found, which will show the true edge line, and is drawn as follows: Take the girth of 6, 7, 8 in the side elevation and place it as shown by the two points 6-7 and 8 on the center line *a b* in the front elevation. Through these two points at right angles to *a b* draw lines and intersect these by lines drawn from similar numbers 6, 7 and 8 in the front elevation. A line traced as shown from 6 to 7′ to 8′ to 8 will give the developed section on 6-7, 8 in the side elevation.

The pattern is developed in J, which shows the half pattern for the dash hood. As 1 8 and 5 6 in the side elevation show the true length of 1 8 and 5 6 in the front elevation, take the length of 1-8 in the side elevation and place it as shown by 1 8′ in J. Using 8′ as center and 8′-2 and 8′-3 in the true lengths as radii, describe the arc 2 and 3 in J. Set the dividers equal to one of the equal spaces in 1-5 in the front elevation, and starting from 1 in J, step to arc 2 and 3 and draw a line from 3 to 8′. Now with 8′, 7′ in the developed section as radius and 8′ in J as center draw the arc 7′, which intersect by an arc, struck from 3 as center and 3, 7′ in the true lengths as radius. With radii equal to 7′ 4 and 7′ 5 in the true lengths, and 7′ in J as center, describe the arcs 4 and 5. Intersect these by stepping with the dividers, from arc 3 to 4 and from 4 to 5, setting the dividers equal to the spaces in 1-5 in the front elevation. Using as radius 7′ 6, in the developed section and 7′ in J as center, describe the arc 6, which intersect by an arc struck from 5 as center and 5 6 in the side elevation as radius. Trace a line through points thus obtained in J, then 1, 5, 6, 8′ will be the half pattern for the dash hood.

Take a tracing of the straight side of the body A B 1, 8, C D E, and place it as shown by A^1 B^1 1′, 8′, C^1 D^1 E^1, and on the line 1′, 8′ place a reproduction of the half dash hood J as shown by 1′, 5′, 6′, 8′ in the pattern. Then A^1 B^1 5′ 6′ C^1 E^1 will be the combined pattern of body and dash, to which riveting flanges must be allowed.

The pattern for the rear panel A E is cut as long at *c c′* in the front elevation and in width equal to the girth of A E. Sometimes a hinged and clasped door is placed in the rear panel, as indicated in diagram H.

PATTERN FOR BODY AND DASH HOOD OF A ROADSTER, THE SIDES OF THE BODY BEING STRAIGHT WITH TAPERING DASH HOOD, THE HOOD BEING DEVELOPED BY RADIAL LINES

Fig. 41 shows a view of a roadster having a body with straight sides combined with a tapering dash hood, in which the radial line method can be employed in developing the pattern. In the previous dash hoods the only taper shown was at the top, the sides being straight. In this problem the taper takes place on the sides as well as the top, which makes a plan view necessary, so as to give a clearer understanding, and is shown in Fig. 42. In this case the taper is uniform, thus allowing the pattern to be obtained from the frustum of a right cone. If, however, the taper were dissimilar at top and sides of the hood, then the pattern would have to be developed by triangulation.

Let *A B C* 1′ 1″ 5″ *D E* represent the side elevation of the body and dash hood; the front elevation of the radiator, as well as the outline of the dashboard and the lower part of the hood being shown placed in its proper position. On the center line *b c* the center point *a* is used in describing the semicircle in the dashboard 5 1 5. Through the center, *a*, draw a line indefinitely, to the right as far as *d* and to the left until it intersects the dashboard at 5 5′ and the side of the body at 5″. Through 5″ draw the vertical line 5^x 5° and extend the line of the taper of the top of the hood 1″ 1′ until it meets the vertical line 5^x 5° at 5° and the line 5″ *d* at *F*. Then 5″ *F* represents the center line of the right cone, one-half elevation of which is shown by *F* 5° 5″. From this the hood 5′ 1′ 1″ 5″ will be cut.

As 5 1 5 in the front elevation shows, the semicircular section on the line 1′ 5′ in the side elevation, then divide one-half of this section or the quadrant 1 5 into equal spaces, as shown. From these intersections, 1, 2, 3, 4 and 5, draw horizontal lines until they intersect the lower line of the hood 1′ C in the side elevation at 1′, 2′, 3′, 4′ and 5′. Through the points 2′, 3′ and 4′ draw radial lines towards the center *F*, extending the lines to the left until they cut the curved outline of the dash hood at 2″ 3″ and 4″. From these points lines are drawn perpendicular to the center line, 5″ *F* until they intersect the side of the cone at 2°, 3° and 4°.

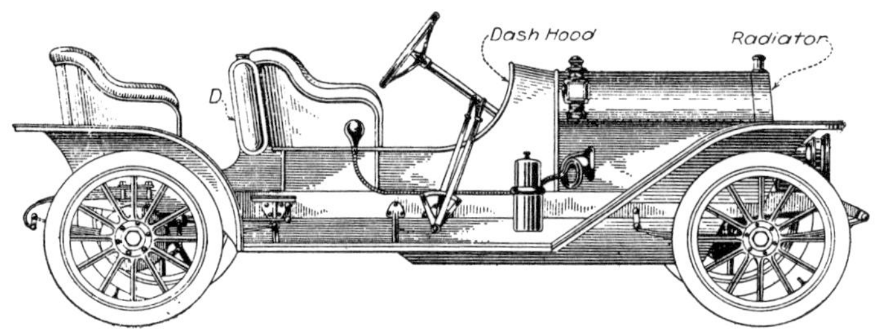

Fig. 41. View of a Roadster

The pattern is now laid out as follows: Using *F* as a center, and *F* 1′ as a radius, draw the arc 5^a 5. Set the dividers equal to one of the spaces in the arc 1 5 in the front elevation, and, starting from 5^a, step off the required number of spaces as shown from 5^a to 1 to 5. Through these points draw lines to the center, *F*, extending them to the left indefintely, as shown, and intersect them by arcs struck from the center, *F*, drawn from similar points of intersections 1″ 2°, 3°, 4° and 5°. Trace a line through points thus obtained and then 5 5^v 1^v 5^v 5^a will be the full pattern for the tapering dash hood.

To show how much of the body will receive the taper of the hood, a half plan is shown below the side elevation, and is drawn as follows: Draw any horizontal line, as *L K*, and from *C* and *A* in the side elevation, project vertical lines as shown, making *K J* equal to one-half of the width of the dashboard, *f i* in the front elevation. As the hood is cut from a right cone, then the taper is equal on all three sides of the hood, as indicated by the distance *e* in the side elevation, between the points 1′ and 5°, as shown by the horizontal dotted lines.

Therefore from *J* in the half plan draw a line parallel to *K L* until it is intersected at *m* by a vertical line dropped from 5″ in side elevation. Take the distance *e* in the side elevation and set it off as shown by *e′* in plan or from *m* to *H* and draw a line from *J* to *H*. From *H* parallel to *K L* draw the line *H G*, meeting the line *L A* at *G*, which completes the half plan. Then *H* in plan, or the line 5″, 5^x in side elevation, shows where a slight bend is made in the side of the body. In making this bend, a sharp corner should be avoided, making a curved sweep, similar to *X Y*.

The combined pattern for body and hood is laid out as follows: Take a reproduction of $A\ B\ 5^x\ 5''\ D\ E$ and place it as shown by $A^1\ B^1\ 5^x\ 5''\ D^1\ E^1$. Extend $B^1\ 5^x$ as $B^1\ J^1$, making the length $5^x\ J^1$ equal to $H\ J$ in plan or $5^v\ 5^a$ in the hood pattern, both being equal. From $5''$, in the body pattern, draw the line $5''\ 5$ parallel and equal to $5^x\ J^1$ and draw a line from J^1 to 5. Then on the line $5''\ 5$ place a reproduction of the half pattern of the hood, 5^v, 1^v, 1, 5^a, as shown by $5''$, 1^v, 1, 5 in the combined pattern. Then $A^1\ B^1\ J^1$ 1, 1^v, $D^1\ A^1$ will be the pattern for the half body and dash hood combined, to which riveting flanges must be added. The curve shown by D^1 represents the tank rest for the gasoline tank shown by D in Fig. 41. Sometimes this same shape of dash hood just described has a different outline, which can also be developed by radial line development.

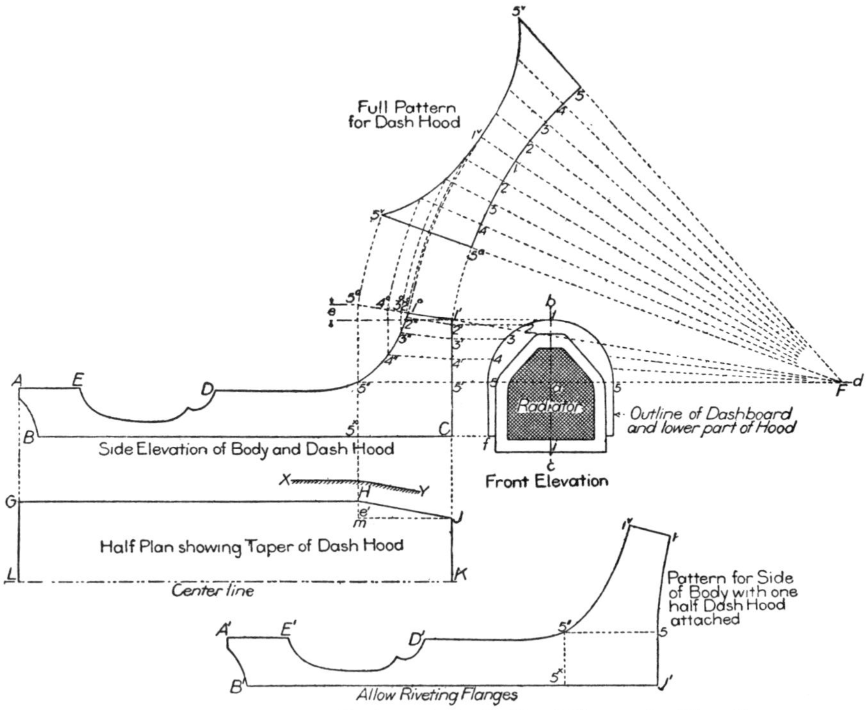

Fig. 42. Pattern for Body and Dash Hood Shown in Fig. 41

PATTERN FOR A DIFFERENT OUTLINE OF CONICAL DASH HOOD

Fig. 43 shows part of the body of the car with a change in the outline of the dash hood, showing how the hood can be laid out by using the same method of development as in the previous figure. Thus D shows the side elevation of the dash hood. A double curved outline is indicated by $1°\ 4''\ 5''$. In its proper position, draw the front elevation of the radiator, from which the outline of the dash hood is obtained, the semicircle being struck from the center b. Through the center b draw the horizontal line b c, extending it to the left until it intersects the dash hood outline at $5''$, extend the pitch of the dash 1° $1'$ until it intersects the center line b c.

Divide one-half of the semicircle in the front elevation from 1 to 5 into equal spaces, from which points draw lines until they intersect the line of the dash hood $1'$ a in the side elevation, as shown from $1'$ to $5'$. Through these points $2'$, $3'$ and $4'$, draw radial lines, extending them to the left until they cut the irregular curve at $2''$, $3''$ and $4''$. Extend the pitch of the hood $1'$, 1° as $1'$ 5°, intersecting same by a vertical line erected from $5''$ at 5°. From the intersections $2''$, $3''$ and $4''$ draw lines at right angles to the center line $5''$ c until they intersect the flare of the cone at 2°, 3° and 4°, respectively. Describe the arc 5, 1, 5, upon which place the girth of twice the amount of spaces contained in the quarter circle in the front elevation. Through these points draw radial lines, extending them upward indefinitely, as shown, and intersecting them by arcs struck, from similar numbered intersections on the pitch line $1'$ 5° in the side elevation, thus obtaining the points of intersections 5^{v} 4^{v} 3^{v} 2^{v} 1^{v} on both sides in the pattern. Trace a line through points thus obtained, as shown.

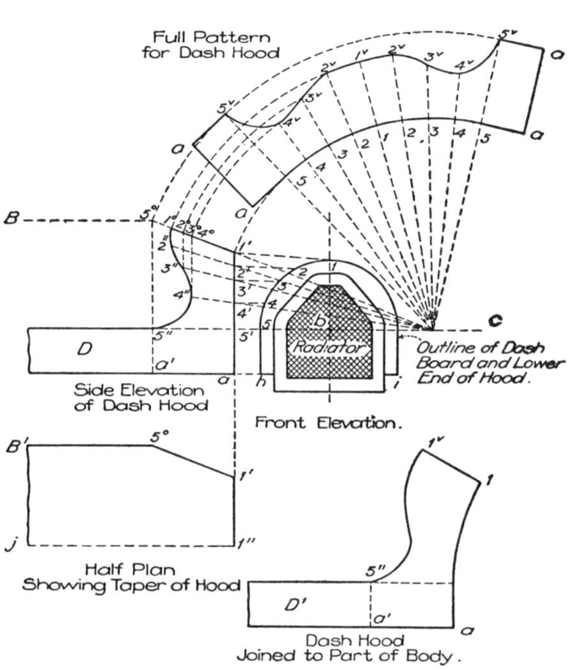

Fig. 43. Conical Hood of Different Outline

At right angles to 5 5^{v} on both sides of the hood pattern, draw 5 a and 5^{v} a equal to $5'$ a or $5''$ a' in the side elevation, and draw the lines a a in the pattern. Then a 5^{v} 1^{v} 5^{v} a a 5, 1, 5, a will be the full pattern for the hood.

As this hood will have a flare on the three sides, the angle of the flare is determined by drawing the horizontal line B from point 5° in the side elevation, then B 5° $1'$, is the desired angle. The half plan showing the taper of the hood is obtained by taking a reproduction of the angle B 5° $1'$ and placing it as shown by B^1 5° $1'$ in the plan and making the vertical line $1'$ $1''$ equal to one-half of the width of the dashboard h i in the front elevation. From $1''$ in the half plan, draw the center line $1''$ j parallel to B^1 5°. The combined pattern of the dash hood joined to part of body is shown by D^1 $5''$ a', which is a reproduction of D $5''$ a' in the side elevation. On the line 5″ a' in the combined pattern place a tracing of one-half the dash hood shown by a, 5^v, 1^v, 1, a. Place the line 5^v a of the hood pattern on the line 5″ a' in D^1 as shown by 5″, 1^v, 1, a, a' 5″, which completes the pattern.

PATTERN FOR BODY AND DASH HOOD COMBINED OF SEMI-RACER, WHOSE SIDES ARE STRAIGHT AND WHOSE HOOD IS DEVELOPED BY TRIANGULATION

In Fig. 44 is shown a view of a semi-racer car, having body and dash hood combined in one. The shape of the hood at either end is such that triangulation is necessary in the development of the pattern, and is explained in detail in Fig. 45, in which A B C 1, 7, 4′ D, shows the side elevation of the body and the hood. In its proper position draw the front elevation of the dash hood, a b c representing the outline of the radiator, d 1 f the vertical outline of the lower part of the hood and 4, 7, 4^x the vertical outline of the upper part of the hood on the lines 1 C and 7, 4′ respectively in the side elevation. Through the center of the front elevation draw the center line E F, and divide the one-half lower outline into equal spaces as shown from 1 to 4 or 3 spaces. In similar manner divide the upper half outline 7 to 4 into the same number of parts as shown by 5, 6 and 7. In this case but 3 spaces have been used in each outline to avoid a confusion of lines. In practice more spaces should be employed.

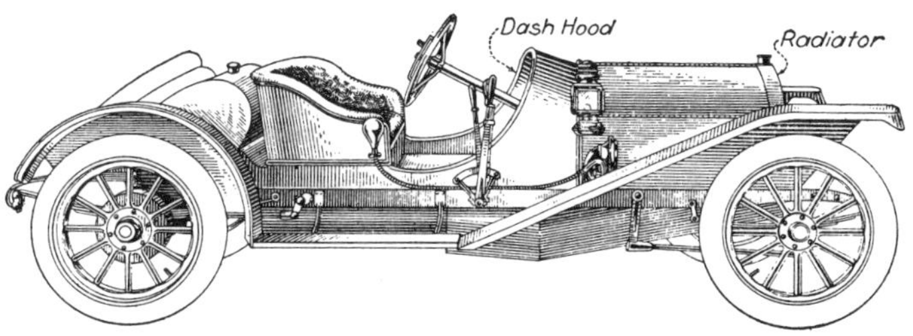

Fig. 44. View of a Semi-Racer

From the divisions 1 to 4 in the front elevation project lines horizontally until they cut the lower line of the dash hood in the side elevation, also from 1 to 4. In a similar manner project horizontal lines from points 4 to 7 in the front elevation until they intersect the upper line of the hood in the side elevation at 4′, 5, 6 and 7. Connect lines in the front elevation as shown from 5 to 4, 5 to 3, 5 to 2, 2 to 6 and 6 to 1; also connect similar numbers in the side elevation as shown. The lines in the front elevation will then represent the bases of triangles the altitudes of which will be equal to the horizontal distances between similar numbered points in the side elevation, and the triangles are obtained as follows: Extend the line C 1 in the side elevation as shown by G H.

Now take the various distances in the front elevation from 1 to 6, 6 to 2, 2 to 5, 5 to 3 and 5 to 4, and place them on the line G H as shown by similar numbers. All points numbered 1, 2, 3 and 4 remain upon the line G H, while from points 5 and 6, lines are projected at right angles to H G and are intersected at 5^v and 6^v by lines drawn parallel to G H from points 5 and 6 on the curve in the side elevation. Draw lines in the diagram of true lengths, from 1 to 6^v, 6^v to 2, 2 to

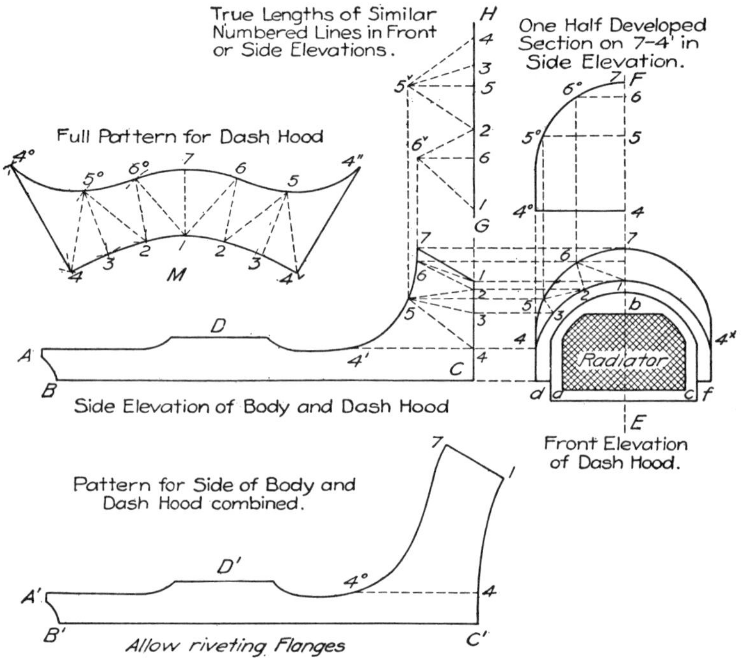

Fig. 45. Pattern for Body and Hood of Car Shown in Fig. 44.

5^v, 5^v to 3 and 5^v to 4, which show the true lengths of similar numbered lines shown in either elevation.

Before the pattern can be completed the true edge line on the curve of the dash hood 7—4′ must be developed as follows: Take the girth of 4′ to 5 to 6 to 7 in the side elevation, and place it as shown by similar numbers on the center line E F in the front elevation. From these points at right angles to F E draw lines which intersect by lines drawn from similar numbers in the semi-profile 4—7 in the front elevation, parallel to F E, thus obtaining intersections 4°, 5°, 6° and 7. A line traced through points thus obtained will give the one-half developed section on 7—4′ in the side elevation. The true lengths on the line 7—1 and point 4 in the front elevation are shown respectively by 7—1 and 4—4′ in the side elevation.

The pattern for the dash hood is now in order. Take the distance of 1—7 in the side elevation and place it on any vertical line in M, as 1—7. With 7—6° in the developed section as radius, and 7 in M as center, describe the arc 6°, which intersect by an arc struck from 1 as center and 1, 6^v in the true lengths as radius. Using 1—2 in the front elevation as radius, and 1 in M as center, describe the arc 2, which intersect by an arc struck from 6° as center and 6^v, 2 in the true lengths as radius. Proceed in this manner, using alternately, first the proper division in the developed section, then the proper true length in the diagram G H; the proper division in the front elevation on the outline *d* 1 *f*, then the proper true length in the diagram H G, until the line 4, 4° in M is drawn, which is obtained from 4 4′ in the side elevation. Trace a line through points thus obtained in M, then 4°, 7, 4″, 4^v, 4 will be the full pattern for the dash hood.

The pattern for side of the body and dash hood combined is obtained by taking a tracing of A B C 4 4′ D in the side elevation, and placing it as shown by A^1 B^1 C^1 4, 4° D^1. On the line 4, 4° place a tracing of the half dash hood 4, 4°, 7, 1 in M, as shown by similar number in the body pattern, which completes the pattern for one side of the body and half dash hood, to which riveting and nailing flanges must be allowed.

PATTERN FOR BODY AND DASH HOOD OF A ROADSTER TYPE, WHOSE BODY SIDES ARE STRAIGHT AND WHOSE HOOD WILL BE DEVELOPED BY TRIANGULATION

Another type of roadster car, the body and dash hood of which are joined in one, is shown in Fig. 46. The body sides are straight and the hood requires triangulation in its solution, as will be explained in Fig. 47, in which *A B C* 1, 8, 4, *D* shows the side elevation of the body and hood, the front elevation of the radiator and dash hood being shown in its proper position as shown by the dotted lines. Note the shape of the radiator, the true semi-profile of the lower end of the dash hood 1, 2, 3, and the semi-vertical section of the upper end of the dash hood 8, 5, 3. In other words this dash hood represents a problem of a transition piece, the lower end of which has straight sides and the upper end an elliptical section.

Through the center of the front elevation draw the center line *E F* and parallel to the outline of the radiator draw the outline of the lower part of the dash hood *a* 1 *b* which will represent the true section on the line 1 *C* in the side elevation. Establish the flare of the hood 1, 8 in the side elevation, which point, 8, project to the center line *F E* in the front elevation and also obtain the point 8.

Establish the point 4 in the side elevation where the curve of the hood is to begin and project this point to the front elevation and obtain a similar number 4 and draw the vertical line 3 4. Then through the points 4, 8, 4^v draw the semi-elliptical section as shown, which will represent the vertical section on the line 4, 8 in the side elevation. Number the divisions in the lower part of the dash hood in the front elevation, 1, 2 and 3, and divide the semi-curve in the upper part of the hood into equal spaces, shown by 4, 5, 6, 7 and 8. Draw lines from 2 to 5, 6, 7 and 8, and from 3 to 5. Project the points 1 to 3 and 4 to 8 to the side elevation, as shown in their proper positions by similar numbers. Also connect lines from 2 to 5, 6, 7 and 8, and from 3 to 4 and 5. The lines in the front elevation will then represent the bases of triangles which will be constructed, the altitudes or hights of which are equal to the horizontal distances between similar numbers in the side elevation, and are obtained as follows: On the line *C H*, extended as *H J*, place the lengths of 3 to 5, 2 to 5, 2 to 6, 2 to 7 and 2 to 8 in the front elevation as shown by similar numbers on *H J*. From the various points 5, 6, 7 and 8 at right angles to *H J* draw lines, which intersect by lines drawn from similar numbers on the curve 4—8 in the side elevation parallel to *C H*, thus obtaining the intersections 5^v, 6^v, 7^v, and 8^v, from which draw lines to 2, and from 5^v draw a line to 3. These slant lines then represent the true lengths of similar numbered lines in the front or side elevation. Next find the true section on the line 4—8 in the side elevation by taking the various divisions on this line and placing them on the center line *E F*, extended in the front elevation as *F G*, as shown by similar numbers. From these small figures at right angles to *F G* draw lines, which intersect by lines drawn from similar numbers in the front elevation, parallel to *F G*, thus obtaining the points of intersections 4° to 7°. The one-half developed section on the line 8—4 in side elevation is then represented by 4, 4° 6°, 8, and these give the true edge line in developing the pattern.

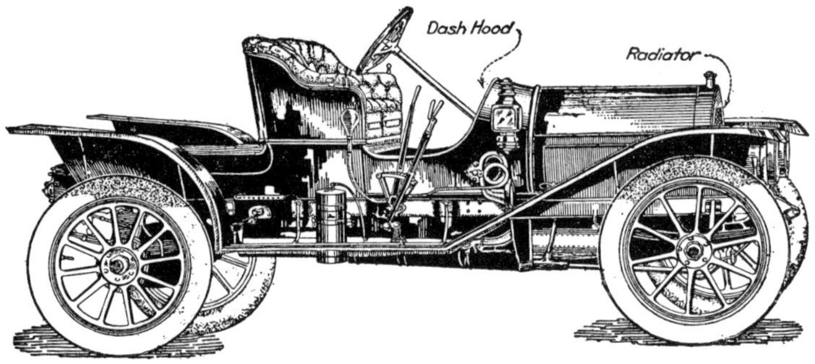

Fig. 46. General View of One Type of Roadster

The pattern for the dash hood is now in order. Take the distance of 1 8 in the side elevation and place it as shown by 1 8 in *K*. With 1 2 in the front elevation as a radius and 1 in *K* as center, describe the arc 2, which intersect by an arc struck with 8 as center and 8^v, 2 in the true lengths as radius. Using 2, 7^v; 2, 6^v; and 2, 5^v in the true lengths as radii and 2 in K as center describe the short arcs 7, 6 and 5. Set the dividers equal to the spaces in the developed section, being careful to measure each space separately, as they are all unequal, as from 8 to 7° to 6° to 5°, and starting from 8 in the pattern K, step to arc 7, then to 6 and 5, using the proper division, and draw lines from 5, 6 and 7 to 2. With 2 3 in the front elevation as radius and 2 in K as center draw the arc 3, which intersect by an arc struck with 5 as center and 5^v 3 in the true lengths as

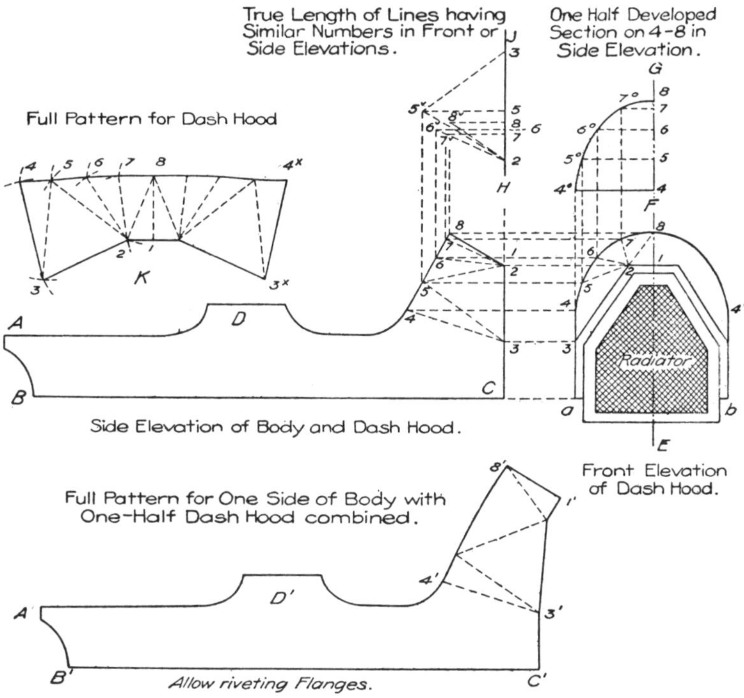

Fig. 47. Pattern for Body and Hood of Car Shown in Fig. 46

radius. Using 3 4 in the side elevation as radius and 3 in K as center describe the arc 4, which intersect by an arc struck with 5 as center, and 5°-4° in the developed section as radius. Trace a line through points thus obtained. Then 1 2 3 4 8 in K will represent the half pattern for the dash hood.

Trace this half pattern opposite the line 1, 8 as shown by 3^x 4^x. Then 4 8 4^x 3^x 1 3 will be the full pattern for the dash hood. Take a tracing of A B C 3 4 D in the side elevation and place it as shown by A^1 B^1 C^1 3′ 4′ D^1. Take a reproduction of 3 4 8 1 in K and place it on the pattern side of body on the line 3′ 4′, as shown by 3′ 4′ 8′ 1′. This then completes the full pattern for body and dash hood combined for one side, to which edges or flanges must be allowed for riveting and nailing to the ash frames.

PATTERN FOR BODY AND DASH HOOD COMBINED OF A RUNABOUT WHOSE SIDES ARE STRAIGHT, THE DASH HOOD BEING DEVELOPED BY TRIANGULATION

Fig. 48 shows a view of a runabout, having a body with straight sides to which the dash hood is connected. The hood requires triangulation in its development and is explained in connection with Fig. 49. Note the peculiar shape of this dash hood as shown in the front elevation. The side elevation of the body and hood combined, with the front elevation drawn in its proper relation to the side, is shown by A B 2 5 C. In this front elevation *a* 3 2 1 *b* represents the true shape of the lower part of the dash hood on the vertical line 1—B in the side elevation, and 4 5 6 4° in the front elevation represents the vertical section on the line 4 5 6 in the side elevation. Connect points in both views by drawing lines from 6 to 2 to 5 to 3. These lines in the front elevation represent the bases of the triangles which will be constructed, the altitudes of which are equal to the horizontal distances shown between similar numbers in the side elevation.

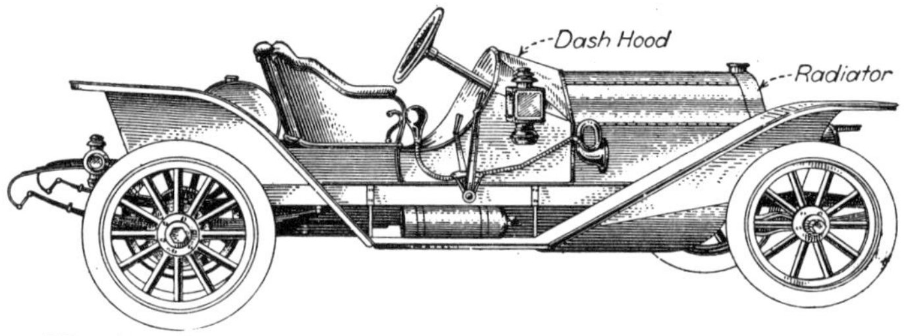

Fig. 48. Runabout with Dash Hood and Body Combined

To find the true lengths in these triangles proceed as follows: Take the lengths of 2 6, 2 5 and 5 3 in the front elevation and place them on the line B G extended in the side elevation as G H, as shown by similar numbers. From the point 5 6, at right angles to H G, draw the line as shown, which intersect by a line drawn parallel to G H from 5 6 in the side elevation, thus obtaining 5^v 6^v in the true lengths. Then draw lines from 5^v to 3, 5^v to 2 and 6^v to 2, which show the true lengths of similar numbered lines in both elevations. The true length of 4 5 in the side elevation is found by placing this distance on the center line D E in the front elevation extended, as E F, as shown by the numbers 4 6 5. Through these points, at right angles to E F, draw lines, which intersect by lines, drawn parallel to E F from similar numbers in the front elevation, thus obtaining points 4° and 5° and 4 4° 5° 6 then represents the one-half developed section on 4 5 in the side elevation.

The pattern is developed as follows: As 1 6 and 3 4 in the side elevation show their true length, then take the distance 1 6 and place it in J as shown by 1 6. With radius equal to 1 2 in the front elevation and 1 in J as center, describe the arc 2, which intersect by an arc struck from 6 as center and 6^v 2 in the true lengths as radius. With 6 5° in the developed section as radius, and 6 in J as center, describe the arc 5, which intersect by an arc struck from 2 as center and 5^v 2 in the true length as radius. With 2 3 in the front elevation as radius, and 2 in J as center, describe the arc 3, which intersect by an arc, struck from 5 as center and 5^v 3 in the true lengths as radius. With 5° 4° in the developed section as radius and 5 in the pattern J as center, describe the arc 4, which intersect by an arc struck from 3 as center and 3 4 in the side elevation as radius. Connect lines to points thus obtained in J, then 1 2 3 4 5 and 6 will be the half pattern for the dash hood.

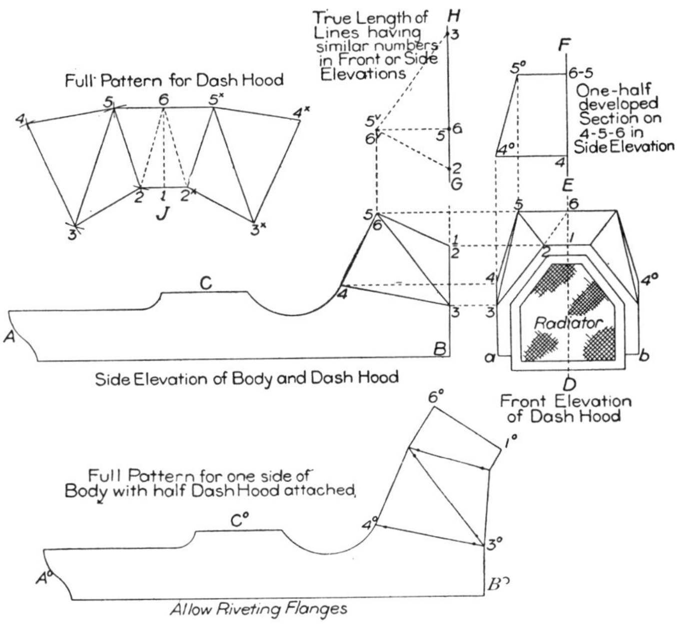

Fig. 49 Pattern for Body and Hood of Car Shown in Fig. 48

To complete the full pattern trace this half opposite the line 1 6, as shown by similar numbers on the right. For the full pattern for one side of the body with half dash hood attached, take a tracing of A B 3 4 C in the side elevation and place it as shown by A° B° 3° 4° C° and on the line 3° 4° add the half pattern of the dash hood 3 4 6 1 in J, as shown by 3° 4° 6° 1° in the pattern, which completes same. Allow riveting and nailing flanges in the same way as explained in a previous problem.

PATTERN FOR BODY AND DASH HOOD OF A ROADSTER HAVING STRAIGHT SIDES, THE HOOD BEING DEVELOPED BY PARALLEL LINES

In Fig. 50 is shown the style of dash hood used in a French type of roadster automobile, in which the body has straight sides with dash hood combined. In this car the hood has no flare and can, therefore, be developed by parallel lines without change of profiles, as was shown in previous problems. Note that in the

French type the radiator B is almost flush with the dash shield or hood and that the motor hood A sets in the front of the radiator. All this is taken up in detail in Fig. 51, in which A B 6 6^1, C D shows the side elevation of the body and dash hood.

In its proper position draw the front elevation of the dash hood, in which the outline of the radiator as well as the French type of motor hood is shown. Through

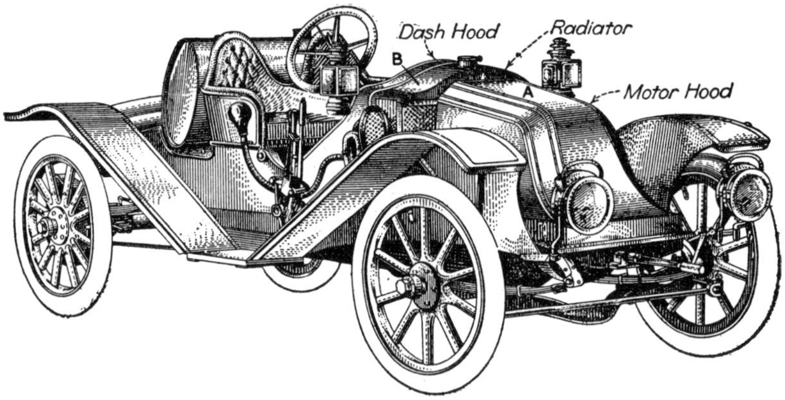

Fig. 50. Roadster with Body and Dash Hood Combined

the center of the front elevation draw the center line F E. As the sides of the body are straight it is only necessary to divide the curved portion of the dash hood into equal spaces, as shown from 1 to 6 in the front elevation, from which points horizontal lines are drawn into the side elevation until they cut the line 6^1 1^1 and

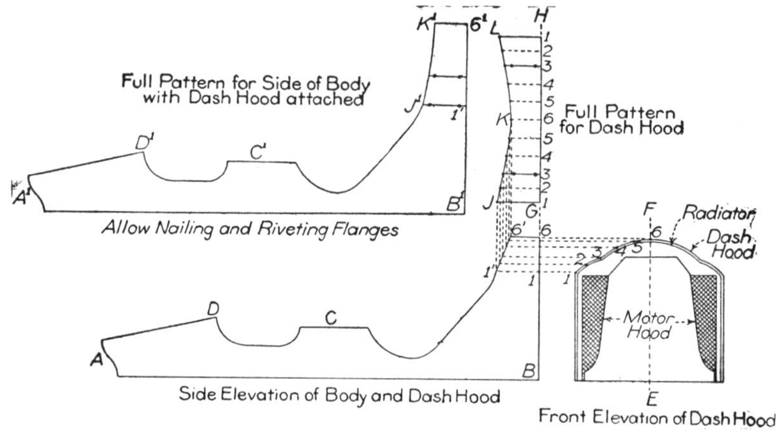

Fig. 51. Pattern for Body and Hood of Car Shown in Fig. 50

1 6, as shown. Extend the line B 6 in the side elevation, as shown by G H, upon which place the girth of the dash hood, as shown from 1 to 6 to 1 on G H. At right angles to G H from the small figures on same draw lines as shown, which intersect by lines drawn parallel to G H, from similar intersections on the hood line 6^1 1^1 in the side elevation. Through points thus obtained trace a line as indi-

cated by J K L, then 1 J K L 1 will be the full pattern for the dash hood or shield.

For the full pattern for the body side with dash hood attached, take a tracing of A B 1 1^1 C D and place it as shown by A^1 B^1 1^1 J^1 C^1 D^1 and on the line J^1 1^1 place a duplicate of the half pattern of the dash hood J 1 6 K, as shown by J^1 1^1 6^1 K^1, which completes the pattern and to which flanges must be allowed for riveting and nailing in the construction work.

PATTERN FOR BODY AND DASH HOOD OF A RUNABOUT, WHOSE SIDES ARE CURVED, THE HOOD HAVING A COMPOUND CURVE AND IS DEVELOPED BY TRIANGULATION

In Fig. 52 is shown a view of a runabout with curved body and dash hood combined. The body combined with the dash hood forms part of a true circle, struck from the center *o* as shown in the front elevation in Fig. 53. In this figure 3‴ 1′ 14 9 8 2 4′ 4″ and 4‴ is the side elevation of the body and dash hood. In its proper position to the right draw the front elevation of the dash hood and body of car, the arc shown from 1 to 8 to 1^v being struck from the center *o* on the center line C D, and represents the vertical section through 2 8 in the side elevation, and the outline shown from 14 to 9 to 14^v representing the vertical section on the line 9 14 in the side elevation.

Fig. 52. Runabout with Curved Body and Dash Hood Combined

As the body is curved, it must first be developed in the flat as follows: Extend the extreme upper edge of the body in the side elevation as 4″ 4′ until it intersects the front elevation at 4. Divide the space between 4 and 1 into equal spaces as shown by 2 and 3; also between 4 and the center point 8 into convenient equal spaces as shown by 5, 6 and 7. From these divisions 1 to 8 draw horizontal lines cutting the outline of the side elevation as shown by the various figures 1 to 8. At pleasure draw any vertical line as H G upon which place the girth of 1 to 4 in the front elevation, as shown by similar numbers on H G. From these small figures at right angles to H G draw lines indefinitely, which intersect by lines drawn parallel to H G from similar numbered intersections in the side elevation as shown. A line traced through points thus obtained as shown by L M 1 2 N O P will be the side pattern for the curved body, to which the dash hood pattern will be joined on the line 2 1.

This dash hood will be developed by triangulation as follows: Divide the half outline of the dash board in the front elevation into as many spaces as there are in the outline of the body between 4 and 8, or four spaces, as shown from 9 to 13.

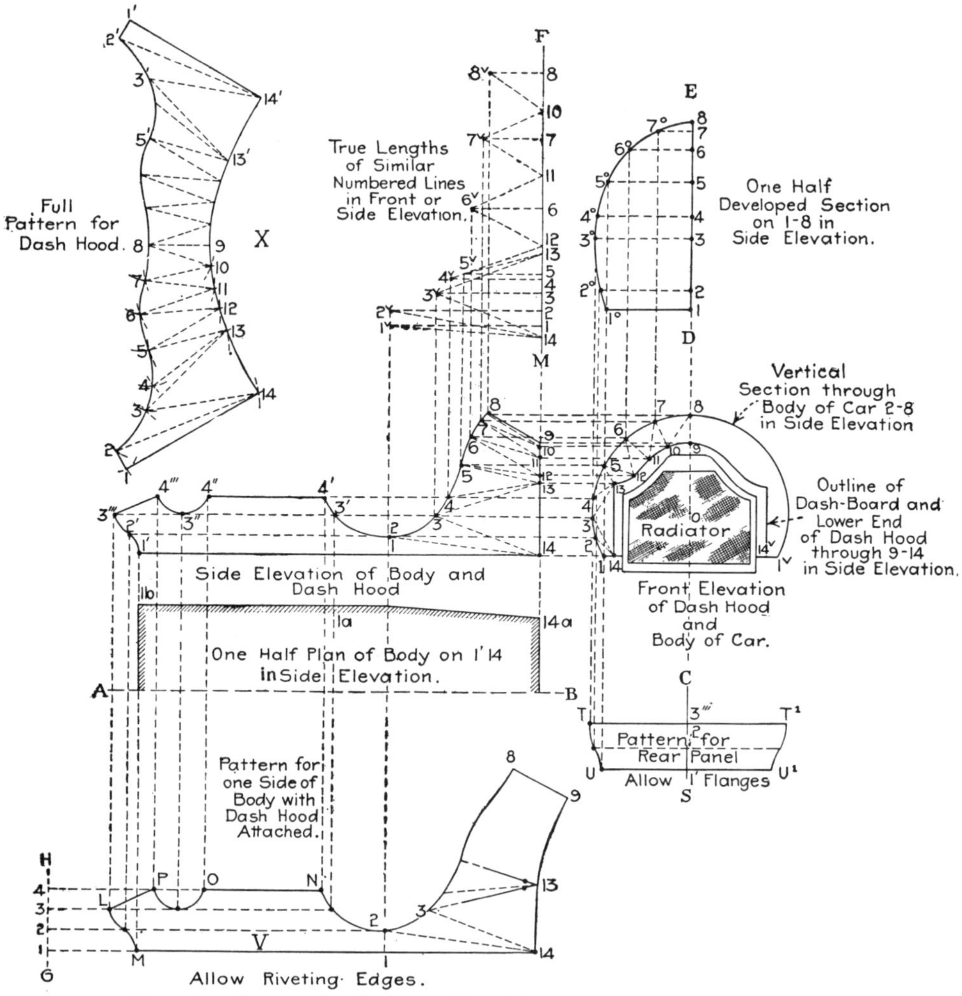

Fig. 53. Pattern for Body and Dash Hood of Car Shown in Fig. 52

From the various divisions 9 to 13 to 14 in the front elevation draw horizontal lines into the side elevation as shown from 9 to 14. Connect lines in the front elevation from 1 2 and 3 to 14; from 3 4 and 5 to 13, and then from 5 to 12 to 6 to 11 to 7 to 10 to 8 to 9. In a similar manner connect similar numbers in the side elevation.

The various lines in the front elevation then represent the bases of triangles which will be constructed, the altitudes of which will equal the horizontal distances between similar numbers shown in the side elevation. Extend the line 14 9 in the side elevation as shown by M F. Take the various lengths in the front elevation, as 14 to 1, 14 to 2, 14 to 3, 13 to 3, 13 to 4, 13 to 5, 5 to 12, 12 to 6, 6 to 11, 11 to 7, 7 to 10 and 10 to 8 and place them on the line M F as shown by similar numbers. At right angles to M F, from the various numbers, excepting 10 to 14 inclusive, draw lines indefinitely, which intersect by lines drawn parallel to F M from similar numbers in the curve of the dash hood in the side elevation, thus obtaining points of intersection from 1^v to 8^v inclusive, and from these numbers draw lines to the proper number on the line F M. For example, as the line connects from 6 to 12 in the front or side elevations, then a line must be drawn in the true lengths from 6^v 12 on the line M F, which shows the true length of 6 12 in the elevations.

The next step is to find the one-half developed section on the line 1 2 8 in the side elevation which will give the true edge line in developing the pattern. Take the girth of 1 2 8 in the side elevation, measuring each space separately, as they are all unequal, and place then on the line C D extended in the front elevation, as shown by similar numbers on D E. At right angles to D E from the small figures draw lines, which intersect by lines drawn parallel to D E from similar numbers 1 to 8 in the outline of the body in the front elevation. Trace a line through points 1° to 8 thus obtained. Then 1 1° 8 will be the one-half developed section on the line 1 8 in the side elevation.

All the true lengths having been found, the pattern is now in order. Draw any horizontal line 8 9 in X equal to 8 9 in the side elevation, which shows its true length. Now with 9 10 in the front elevation as radius and 9 in X as center describe an arc, which intersect by another arc struck from 8 as center and 8^v 10 in the true lengths as radius. Now, with 8 7° in the developed section as radius and 8 in X as center, describe the arc 7, which intersect by another arc struck from 10 as center and 10 7^v in the true lengths as radius. Proceed in this manner, using alternately first the divisions in the dash board in the front elevation, then the proper slant line in the true lengths; the divisions in the one-half developed section, then again the true length of the proper line, shown in the diagram of true lengths, until the line 1 14 in the pattern X has been obtained, which is equal to 1^v 14 in the true lengths.

By following the reference numbers no mistake can be made because the various divisions 9 to 14 in X are obtained from 9 to 14 in the front elevation;

the various divisions 1 to 8 in X are obtained from 1° to 8 in the developed section, while the dotted lines in X are obtained from the true lengths. A line traced through points thus obtained in X, as shown by 1 2 8 9 14, will be the half pattern for the dash hood. If desired this half can be traced opposite the line 8 9 in X, as shown by 8 5′ 3′ 2′ 1′ 14′ 13′ 9, which completes the full pattern for the dash hood.

Take a tracing of 1 2 3 8 9 13 14 in X and place it to the body pattern in V as shown by similar numbers, placing the line 1 2 of the dash X onto the line 1 2 of the body V. Then L M 14 9 8 2 N P will be the pattern for one side with half dash hood combined, to which riveting edges must be allowed.

When the half dash hood is not attached in one to the side of the body, a seam is usually made along 1 2 in the side elevation, and the dash hood is made in one piece as shown in X. This makes the body in four parts, the dash hood, the two sides of the body and the rear panel.

As this rear panel must be developed, owing to the curve of the body, the following method gives the best results. Extend the line E C in the front elevation, as shown by C S. On this line C S place the girth of the rear panel 3‴ 1′ in side elevation, as shown by similar numbers on C S. Through these small figures at right angles to C S draw lines as shown, which intersect by lines drawn parallel to C S from similar numbered intersections in the front elevation. Trace a line through points thus obtained and T U will be the miter cut. Trace 1′ U T 3‴ opposite the line 1′ 3‴ as shown. Then T T^1 U^1 U is the pattern for the rear panel to which flanges must be allowed. One-half plan of the body on 1′ 14 in the side elevation is shown by A 1^b 1^a 14^a B.

PATTERN FOR DASH HOOD AND FRONT PART OF BODY IN A GUNBOAT TYPE OF AUTOMOBILE, THE BODY BEING CURVED AND HOOD BEING DEVELOPED BY TRIANGULATION

In Fig. 54 is shown a perspective view of an automobile with a gunboat body and the dash hood attached to the front part of the body, the body being curved as shown in the front elevation in Fig. 55. In this figure but part of the side elevation of the body is shown, as the body is developed only to the line 5 9. The method of developing the full side of the body when the sides are curved, as shown from 5 to 9 in the front elevation, was shown in the text accompanying Figs. 3,

24 and 53. Thus d 4 1 13 9′ 9 f in Fig. 55, shows the side elevation of the dash hood and part of the body. In its proper position in relation to the side, draw the front elevation h 3 2 2′ 3′ i, showing the true section on the line 1 4 in the side view and representing the outline of the dashboard as well as the lower end of the dash hood, and 5 9 13 9″ 5′ showing the vertical section or outline of the upper end of the hood on the line 13 9′ 9 5 in the side elevation. That part of the curve from 5 to 9 in the front elevation shows the true section through the body on the line 5 9 in side view.

This gunboat type of body and hood presents an interesting problem in triangulation. In the gunboat and torpedo types of automobiles, as well as in the large touring cars, the dash hood is usually made in one piece connecting to the front portion of the body only, or as far as indicated by a in Fig. 54, the rest of the body being made separate.

Fig. 54. Automobile with Gunboat Body and Dash Hood Attached

Through the center of the front elevation in Fig. 55, draw the center line A B and number the lower end of the dash hood, as shown by 1, 2, 3 and 4. In a similar manner space the profile of the body in the front elevation into equal spaces, in this case four, as shown by 5, 6, 7, 8 and 9 and also space the half outline of the upper part of the hood from 9 to 13, as shown by the small figures 9, 10, 11, 12 and 13. From the intersections 1 to 4 in the front elevation, draw horizontal lines to the side view intersecting the proper plane 1 4 as shown. In a similar manner draw horizontal lines from 5 to 13 in the front elevation, until they intersect the line 5 9 13 in the side elevation as shown by similar numbers. Connect lines in both views from 9, 10, 11, 12 and 13 to 2; from 5, 6 and 7 to 4; and from 7, 8 and 9 to 3. These lines in the front elevation then represent the bases of triangles which will be constructed, the altitudes or hights of which are equal to the horizontal distances between similar numbered points in the side elevation; the slant lengths or the hypotenuses of these triangles thus found will give the true length of the lines desired. For example: to find the true length of the line 9 2 in the front elevation, take this distance and set it off on the line 4 1 extended in the side elevation as D E, as shown by the distance from 9 to 2. From 9 erect a perpendicular 9 9^v, which intersect at 9^v by a vertical line erected from 9′ in the side elevation. Draw a line from 9^v to 2 in the true lengths, which will be the true length of similar numbered lines in either elevation. Proceed in this manner for the rest of the true lengths. Take the lengths of the various lines in the front view and place them as shown by similar numbers on D E. Draw horizontal lines from all points on D E, excepting 2, 3 and 4, and intersect them by perpendicular lines erected from similar numbered points in the side elevation as shown in the accompanying Fig. 55.

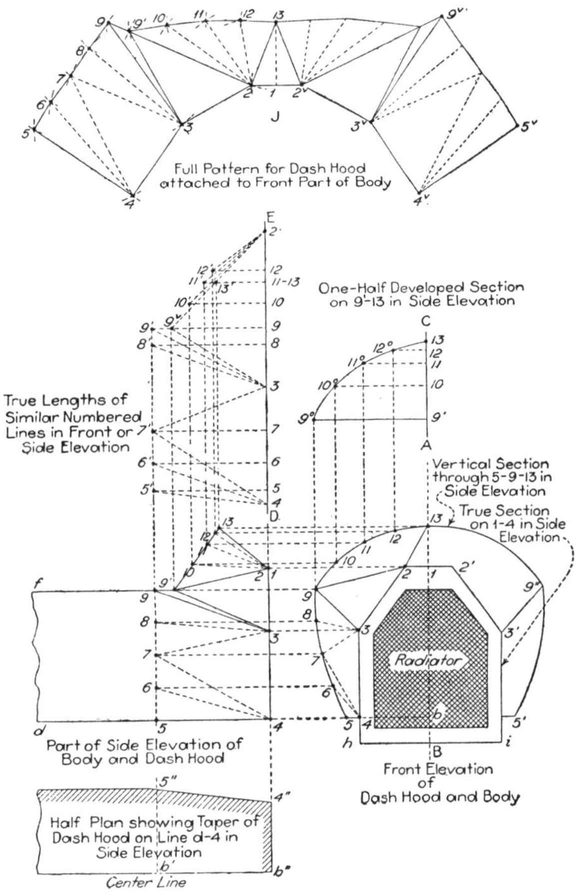

Fig. 55. Pattern for Dash Hood on Car in Fig 54

As the true length of the lines have been found, the true section or edge line on the line 9′ 13 in side elevation must be developed by taking the various intersections on 9′ 13 and placing them on B A extended in the front elevation as A C, as shown by similar numbers, through which at right angles to A C lines are drawn and intersected by lines drawn parallel to A C from similar numbers in 9 13 in the front elevation. A line traced through points thus obtained, as shown by 9° 13 9′, will be the one-half developed section on 13 9′ in the side elevation.

The pattern shape is now in order. As 1 13 in the side elevation shows its true length, take this distance and place it as shown by 1 13 in J. With a radius equal to 1 2 in the front elevation and 1 in J as center, describe the arc 2, which intersect by an arc struck from 13 as center and 13′ 2 in the true lengths as radius. With radii equal to 2 12′, 2 11′ 2 10′ and 2 9^v in the true lengths, and 2 in J as center, describe the arcs 12, 11, 10 and 9′. Set the dividers equal to the various divisions in the developed section from 13 to 12° to 11° to 10° to 9° and starting from 13 in J step to arc 12, 11, 10 and 9′. Using 2 3 in the front elevation as radius and 2 in J as center, describe the arc 3, which intersect by an arc struck from 9′ as center and 9^v 3 in the true lengths as radius. With 9′ 9 in the side elevation as radius and 9′ in J as center describe the arc 9, which intersect by an arc struck from 3 as center and 3 9′ in the true lengths as radius, With radii equal to 3 8′ and 3 7′ in the true lengths and 3 in J as center, describe the arcs 8 and 7, which intersect by divisions obtained from 9 8 and 8 7 in the front elevation as radius. With a radius equal to 3 4 in the front elevation and 3 in J as center, describe the arc 4, which intersect by an arc struck from 7 as center and 7′ 4 in the true lengths as a radius. With radii equal to 4 6′ and 4 5′ in the true lengths and 4 in J as center, describe the arcs 6 and 5, and intersect same with divisions 7 to 6 to 5 in the front elevation as radius. Trace lines in J through points thus obtained and 1 13 9 5 4 3 2 1 will be the half pattern. Trace this half opposite the line 13 1 as shown. Then 5 9 13 9^v 5^v 4^v 1 4 will be the full pattern, with the front part of the body attached, up to 9 5 in the side elevation.

The half plan showing the taper of the dash-hood when connected to the body, is shown below the side elevation. While this half plan is not required in the development of the pattern, it shows a horizontal view on the line *d* 4 in the side elevation.

The distance of *b*″ 4″ in the half plan is obtained from *b* 4 in the front elevation, while *b*′ 5″ in the plan is obtained from *b* 5 in the front elevation. Whether the shape of the body is concave or convex the principle is the same as just shown and explained.

PATTERN FOR DASH HOOD AND FRONT PART OF BODY IN A TORPEDO TYPE OF AUTOMOBILE, THE SIDE OF THE BODY BEING STRAIGHT AND THE DASH HOOD BEING DEVELOPED BY TRIANGULATION

A view of a torpedo model is given in Fig. 56, in which is shown a scuttle dash hood connected to a straight sided body. The pattern for the dash hood is developed by triangulation, which is explained in Fig. 57. In this figure but part of side elevation is shown, because in practical work a joint is usually made at the side of the door. In this case a joint will be made along 6 *b* in the side elevation.

The first step is to draw a correct view of the side elevation, making the curve of the scuttle dash as desired, as shown by A B *c* 1 10 6 A. In its proper position in line with the side, draw the front elevation of the dash hood as shown. Thus 1 5 *c*′ *c*″ 5′ shows the outline of the dashboard and lower end of hood and 10 5 *c*′ *c*″ 5′ 10 the vertical section of the upper end of the dash hood on the line 10 6 *b* in the side elevation.

Draw the center line C D through the front elevation as shown. Extend the top line of the body A 6 in the side elevation until it intersects the outline in the front elevation at 6. Divide the lower outline 1 5 in the front elevation into any number of convenient spaces, in this case four as shown, and in a similar manner divide the upper outline 6 to 10 into a similar number of spaces. Project all points from 1 to 5 horizontally until they intersect the vertical line in the side elevation, as shown by similar numbers. In a similar manner from the small figures 6 to 10 in the front elevation draw horizontal lines intersecting the

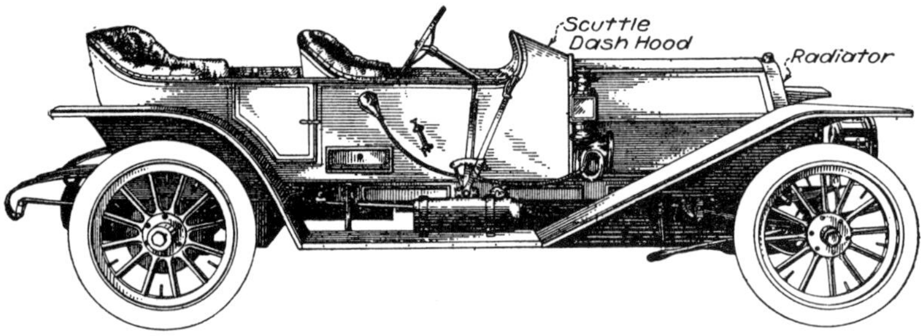

Fig. 56. Automobile with Scuttle Dash Hood Connected to Body

double curve of the dash hood in the side also from 6 to 10. As the curve from 6 to 7 in the side elevation is deep, establish an extra point as *a* which project to the front elevation also obtaining *a*. Connect the various numbers in both front and side elevation as shown by similar numbers. The lines in the front will then represent the bases of triangles, the altitudes of which will equal the horizontal distances between similar numbered points in the side elevation, the hypotenuses of the triangles showing the true lengths.

The methods of obtaining these true lengths are as follows: Extend the line *c* 1 in the side elevation as shown by M F. Take the various distances 1 to 9 to 2 to 8 to 3 to 7 to 4 to *a* and 4 to 6 in the front elevation, and place them as shown by similar numbered spaces on the line M F. All numbers from 1 to 5 on this line M F remain fixed, because all points, 1 to 5, in the side elevation are upon the vertical line 1 *c*. From the various points 6 to 9 on the line M F, draw horizontal lines, which intersect by lines drawn parallel to M F, from similar numbered intersections in the dash curve 6 10 in the side elevation, thus obtaining the intersections 6^v, 7^v, 8^v, 9^v and a^v. Draw slant lines from 1 to 9^v, 9^v to 2, 2 to 8^v, 8^v to 3, 3 to 7^v, 7^v to 4, 4 to a^v and 4 to 6^v, which shows the true lengths of similar numbered lines in front or side elevations.

The true edge line or developed section on the line 6 10 in the side elevation must now be developed as follows: Extend the center line C D in the front elevation as shown by D E, upon which place the girth of the curve 10 6 of the dash hood in the side elevation, as shown by similar numbers on the line D E. At right angles to D E from the small figures 6 to 9 inclusive draw lines, which intersect by lines drawn parallel to D E from similar numbers in the front elevation, thus obtaining the intersections 6° to 9°. A line traced through points thus obtained as shown by 6 10 6° 6 will be the desired pattern and give the true edge line.

As the sides of the body are straight as shown by 5 *c′* in the front elevation, then will *b c* 5 6 in the side elevation show its true shape and will be added to the

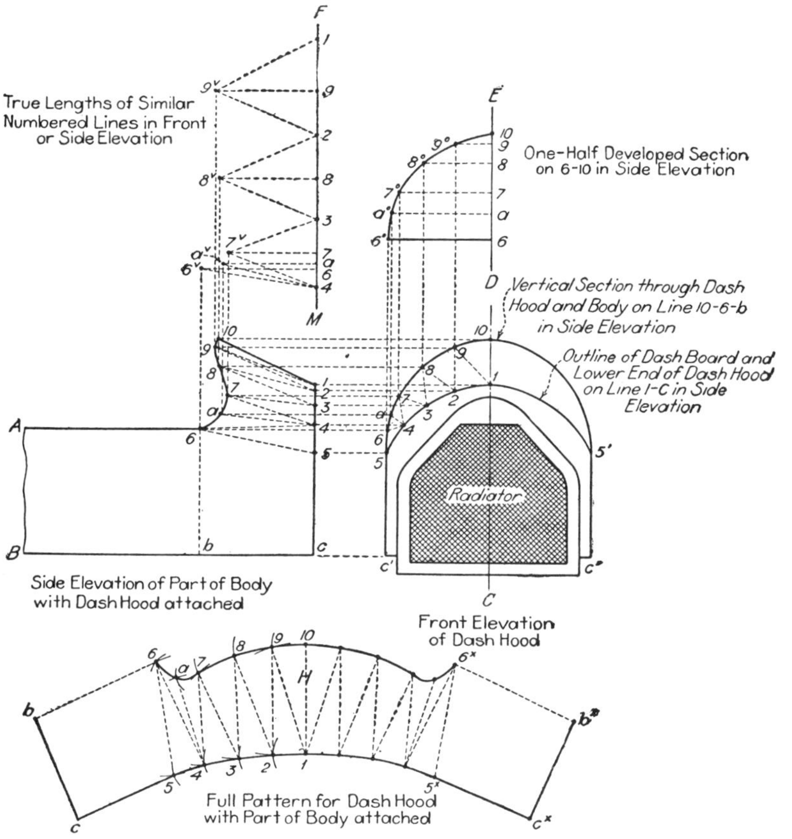

Fig. 57. Pattern for Dash Hood on Car in Fig. 56

pattern for the dash hood, thus leaving the seam or joint along *b* 6 as shown. The joint may be made at any other convenient place where desired.

The pattern for the dash hood is now in order and is developed as follows: Take the distance of 1 10 in the side elevation, which shows its true length, and place it as shown by 1 10 in diagram H. Using 10 9° in the developed section as radius and 10 in H as center draw the arc 9, which intersect by an arc struck from

1 as center and 1 9^v in the true lengths as radius. Using 1 2 in the front elevation as radius and 1 in H as center describe the arc 2, which intersect by an arc struck from 9 as center and 9^v 2 in the true lengths as radius. Proceed in this manner, using alternately, first the divisions in the one-half developed section which shows the true length along 10 6 in the side elevation, then the proper slant line in the true lengths, the divisions along the outline of the dash board in the front elevation, then again the proper true length of the slant lines in the true lengths, until the line 5 6 in the pattern H has been obtained, which is equal to the true length shown by 5 6 in the side elevation. Trace a line through points thus obtained in the pattern H, as shown from 1 to 5 to 6 to 10, and on the line 5 6 trace a reproduction of 5 6 *b* *c* in the side elevation, as shown by similar numbers and letters in the pattern H. This completes the half pattern. Trace this half pattern opposite the center line 10 1 as shown by 1 5^x c^x b^x 6^x 10. Then *b* 6 10 6^x b^x c^x 1 *c* will be the full pattern, to which edges must be allowed for riveting and nailing, when constructing the body.

PATTERN FOR TAPERING DASH HOOD COMBINED WITH BODY IN A RACER CAR THE DASH HOOD JOINING A MOTOR HOOD OF PECULIAR SHAPE, REQUIRING TRIANGULATION IN ITS SOLUTION

In Fig. 58 is shown a type of racer car with a tapering dash hood joining a motor bonnet or hood of the shape shown by 1 6 7 10 in the front elevation in Fig. 59. In this figure A B 10 1 16 11 C D A shows the side elevation of the body and dash hood and 10 7 1 10^e 16 11 10 the front elevation of the dash hood. Note the curve from 7 to 10 at the bottom of the dash hood, which tapers to a given point 10′ in the side elevation. Also note the pitch of the seat base D C in the side view and that the line 10′ 11 is established at pleasure, and when thus established gives the bending line between the body and hood as indicated at 10^c in the half plan.

When the front and side views have been drawn in their relative positions, the first step is to divide the curved parts in the dashboard or motor hood into equal spaces as shown from 1 to 6 and 7 to 10, and from these points draw lines horizontally to the side elevation, intersecting the vertical plane of the dashboard or lower end of the dash hood from 1 to 10. In a similar manner in the front elevation, divide the curved part of the upper end of the dash hood into the same number of spaces as contained in the curves from 1 to 6 or five spaces as shown from 11 to 16.

From these divisions 11 to 16 draw horizontal lines into the side elevation, intersecting the line 10′ 11 16 in the side elevation as shown by similar numbers. Connect points between 1 and 6, and 11 and 16, with lines, as shown in the front elevation, also draw a line from 11 to 7 and from 7, 8 and 9 to 10 (not shown). In a similar manner connect similar points in the side elevation. The connecting lines in the front elevation then represent the bases of triangles which will be constructed, the altitudes or hights of which will be equal to the horizontal distances between similar numbers in the side elevation. Thus the line 10 1 in the side elevation is extended as L K. On this line place the various divisions shown in the front elevation from 1 to 15 to 2 to 14, etc., as shown by similar numbers on the line L K.

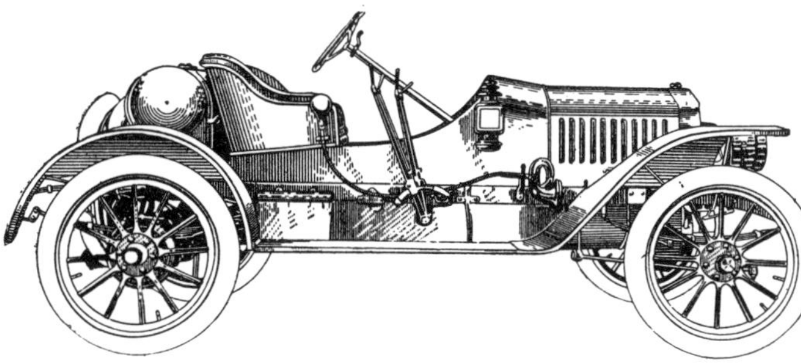

Fig. 58. Racer Car with Tapering Dash Hood Joining Motor Hood

All points from 1 to 10 remain on this line, while all points from 10 to 15 are projected at right angles to L K and intersected by lines drawn parallel to L K from similar numbers on the curve of the dash hood 11 16 in the side elevation, thus obtaining the intersections 10′ to 15′. From these points draw the slant lines to the proper numbers as shown. These slant lines then represent the true lengths of the lines shown by similar numbers in either the front or side elevation.

The next step is to find the true length of the line 11 16 in the side elevation as follows: Take the girth of the various spaces from 11 to 16 in the side elevation and place them on the line E F extended in the front elevation as F G as shown by similar numbers, through which at right angles to F G draw lines, which intersect by lines drawn parallel to F G, from similar numbered intersections in the front elevation, thus obtaining the points of intersections 11° 12° 13° 14° and 15°. A line traced through points thus obtained will be the one-half developed section on 16 11 in the side elevation.

When the pattern for this peculiar dash hood is being developed, the divisions from 1 to 10 in the front elevation are used for the lower end of the hood, and the divisions from 11° to 16 in the developed section for the upper end of the hood while the slant lines in the true lengths are used for the opposite connecting points. As 1 16 in the side elevation shows its true length, take this distance and place it as shown by 1 16 in P. Then with 16 15° in the developed section as radius and 16 in P as center describe the arc 15, which intersect by an arc struck from 1 as center, and 1 15′ in the true lengths as radius. With 1 2 in the front elevation as radius and 1 in P as center describe the arc 2, which intersect by an arc struck from 15 as center and 15′ 2 in the true lengths as radius. Proceed in

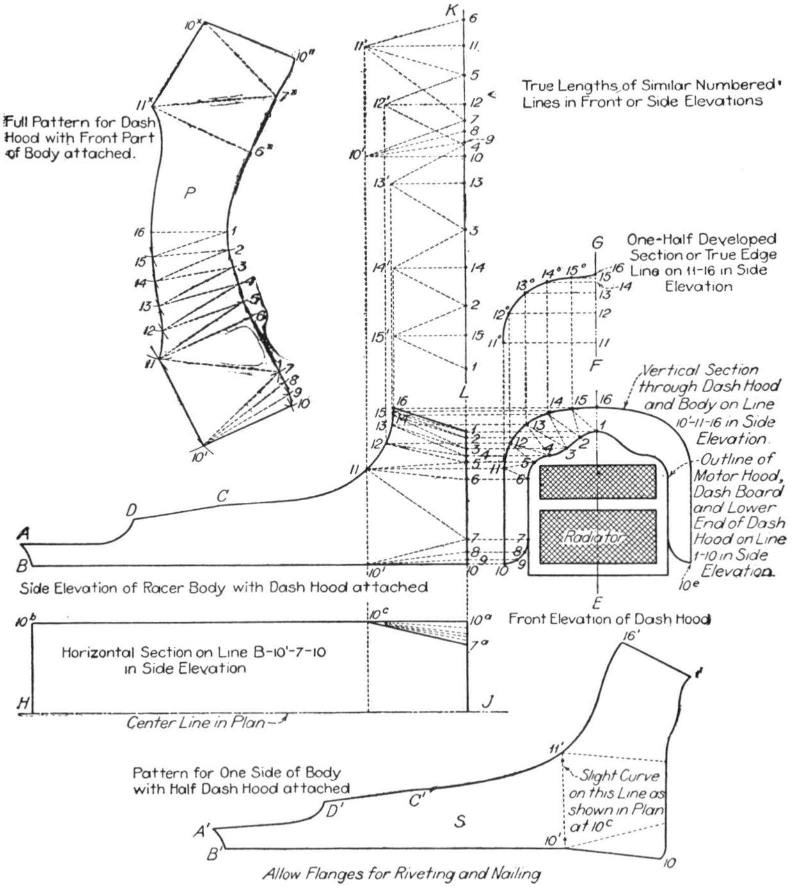

Fig. 59. Pattern for Dash Hood Attached to Body of Car Shown in Fig. 58

this manner, using alternately, first the divisions in the developed section, then the proper radius, in the true lengths, the divisions between 1 and 6 in the front elevation then the proper slant lines in the true lengths, until the line 11 6 in the pattern P has been obtained. Now using 6 7 in the front elevation as radius and 6 in P as center describe the arc 7, which intersect by an arc struck from 11 as center and 11′ 7 in the true lengths as radius. Then using a radius equal to 11 10 in the front elevation and 11 in P as center, describe the arc 10′ which intersect by an arc struck from 7 as center and 10′ 7 in the true lengths as radius.

With radii equal to 10′ 8, 10′ 9 and 10′ 10 in the true lengths and 10′ in P as center draw the arcs 8, 9 and 10. Set the dividers equal to the divisions 7 to 8, 8 to 9 and 9 to 10 in the front elevation and starting from 7 in the pattern P, step to arc 8, then to arc 9 and 10 and draw a line from 10 to 10′. Through points thus obtained trace a line from 1 to 10 and 10′ to 16 which is the half pattern. Trace this half pattern opposite the line 1 16 as shown by 1 6^x 7^x 10″ 10^x 11^x 16. Then will 10^x 10′ 10 10″ be the full pattern for the dash hood.

For the pattern for one side of the body with dash hood attached, take a tracing of A B 10′ 11, C D A and place it as shown in S, by A^1 B^1 10′ 11′ C^1 D^1 A^1. On the line 10′ 11′ place a tracing of the half pattern P, shown by 1 10 10′ 11 16, placing the line 11 10′ on the line 11′ 10′ in S as shown by similar numbers, Then will A^1 B^1 10 1′ 16′ C^1 D^1 be the pattern for the side of body and half dash hood combined, to which riveting edges must be allowed.

This problem completes 16 different types and shapes of dash hoods with body combined which should make the mechanic proficient to develop any other shape that may arise, as the principles just explained may be applied to any form.

METHOD OF FASTENING SHEET METAL DASH HOODS TO WOODEN OR SHEET METAL DASHBOARD

The dashboards are usually made from wood and sometimes from sheet metal. A wooden dashboard placed in its proper position is shown by A in Fig. 60. On this dashboard A, the metal dash hood is fastened in a manner which will be

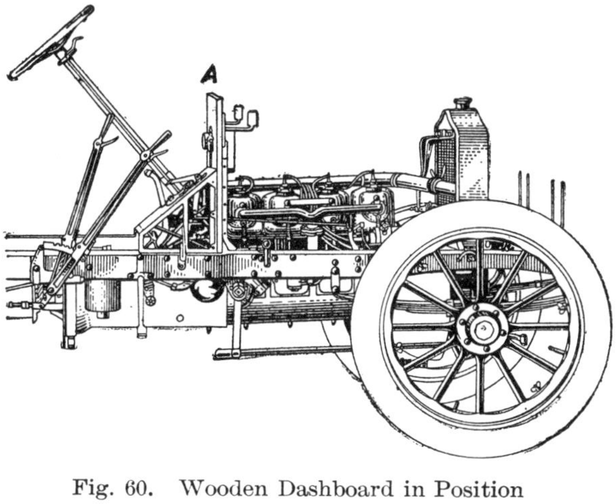

Fig. 60. Wooden Dashboard in Position

Fig. 61. Finishing the Connection between the Metal Dash Hood and the Metal Dashboard

described, and over the joint between the wood and metal a semi-elliptical shaped brass moulding is fastened as indicated by A in Fig. 61. The method employed in making the joint between the metal dash hood and wooden dashboard is explained in detail in Fig. 62. When the metal dash hood is made in two halves with a seam along the top it is riveted every 2 in. along the top as shown in diagram B by *b*, the one side of the hood having an offset flange as shown by *a* which gives a smooth surface along the top. One pound tinned rivets are used and the joint sweated with solder and scraped smooth as previously described.

After the dash hood has been bent and formed correctly to fit the dashboard a half oval polished brass molding is riveted flush with the raw edge of the dash

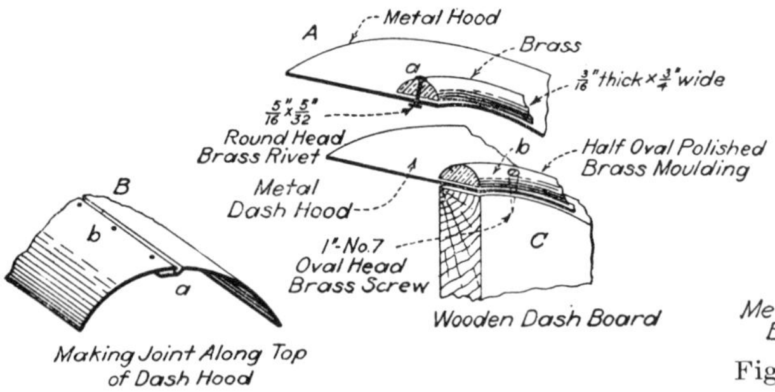

Fig. 62. Joint between Metal Dash Hood and Wooden Dashboard

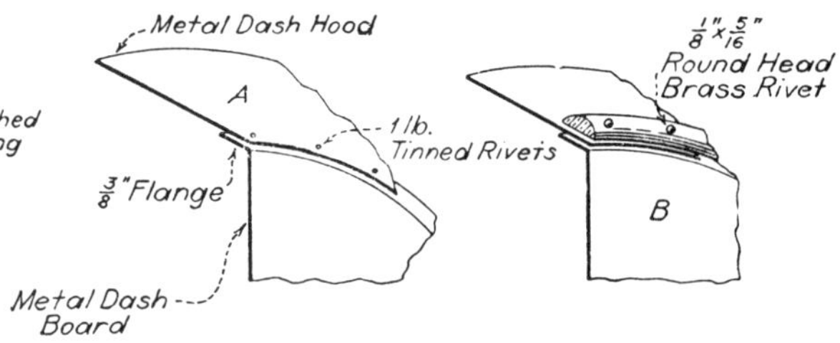

Fig. 63. Method of Joining Hood to Metal Dashboard

hood, as shown in diagram A, the usual size of the mold being 3-16 in. thick × 3-4 in. wide, which is riveted along the front edge of the dash hood with 5-16 × 5-32 in. round head brass rivets placed 4 in. apart, as shown by *a*. After this brass molding is riveted on, the hood is set over the wooden dashboard and 1 in.

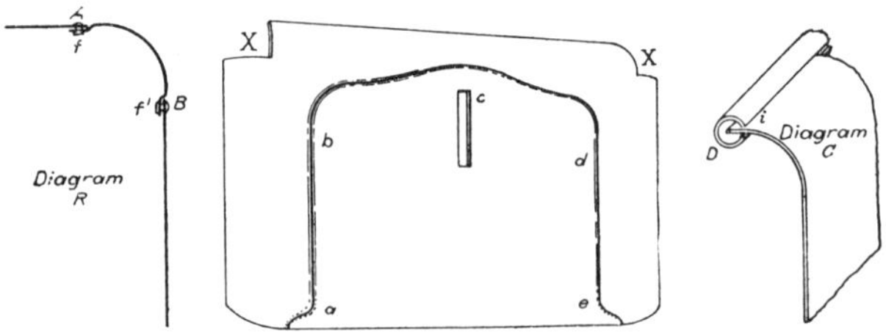

Fig. 64. View of Pressed Metal Shallow Dashboard

Fig. 65. Pressed Dashboard in Position

No. 7 oval headed brass screws are placed midway between the rivet heads in holes previously punched, as shown by *b* in diagram C.

When the dashboard is made from metal as is sometimes the case, the hood is joined to the metal dashboard as shown in Fig. 63. A 3-8 in. flange is turned in on the dashboard as indicated in diagram A, and over this flange the edge of the dash hood runs flush and is riveted with 1 lb. tinned rivets, and then the brass moulding above refered to is put around the top flush with the edge and hiding the rivets. This moulding is riveted to the dash hood every 2 in. with 1-8×5-16 in. round head brass rivets, as shown in diagram B. On many types of cars, where no dash hood is used, shallow dashboards are pressed from sheet aluminum or No. 20 gauge sheet steel, with a curved shield and rounded corners, as shown in Fig. 64, in which at X X openings are allowed to which the rounded corners are riveted and soldered. These pressed rounded corners are in the shape of a quarter ball with an offset flange at A and B in diagram R. These flanges, both on the dashboard and rounded corners, are first cleaned, then tinned and riveted with 1 lb. tinned rivets at *f* and *f′* the metal being slightly countersunk on the outside, so as to have a smooth even surface. The joint is then soldered and scraped clean and smooth. When pressing these dashboards the outline of the motor bonnet or hood is slightly pressed outward, as shown by *a b c d e*, which forms the hood support at the dash end. Around the exposed edge of the dashboard a soft steel black tubing is soldered at *i* as shown in diagram C by D. These black steel tubes are usually made 9-16 in. inside diameter with a 1-32 in. wall, and are soldered to the dashboard in a manner as described in connection with Fig. 29. When this style of metal dashboard is finished and in position it will look as shown by X X in Fig. 65.

DEVELOPING THE PATTERNS AND CONSTRUCTING SHEET METAL SEATS

The next subject of importance in automobile construction is that of laying out and constructing sheet metal seats. In general, seven different types of seats are employed, including a seat for a motor truck, and are known as follows: Two passenger seat, individual seat, single rumble seat, double rumble seat, racer seat, surrey seat and motor truck seat. In laying out the patterns for these seven types parallel lines, radial lines and triangulation systems of developments will be employed and each type will be taken up in the order named in the foregoing.

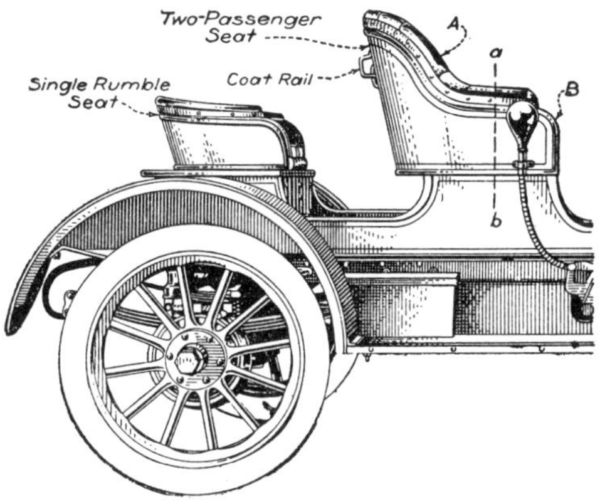

Fig. 66. View of Two Passenger Seat and Single Rumble Seat

The first seat to be developed, will be that of a two passenger seat shown in elevation and plan in the scale drawing in Fig. 2. By referring to H in the side elevation, it will be seen that the seat has a bead or swedge along the bottom sides and top.

In Fig. 66 A shows this same type of seat, placed in position, behind which is shown a single rumble seat, the laying out of which will be taken up in regular order. This cut A also shows the bead or swedge, which gives a neat finish to the car. How the seats are beaded in the flat, then formed up and flanged, and then secured to the ash frame of the seat as well as to the body, will be explained in regular sequence.

Pattern for Two Passenger Seat

The first step is to draw a correct side elevation of the two passenger seat required, as shown by 2 7 8 14 in Fig. 67, below which is drawn in its proper

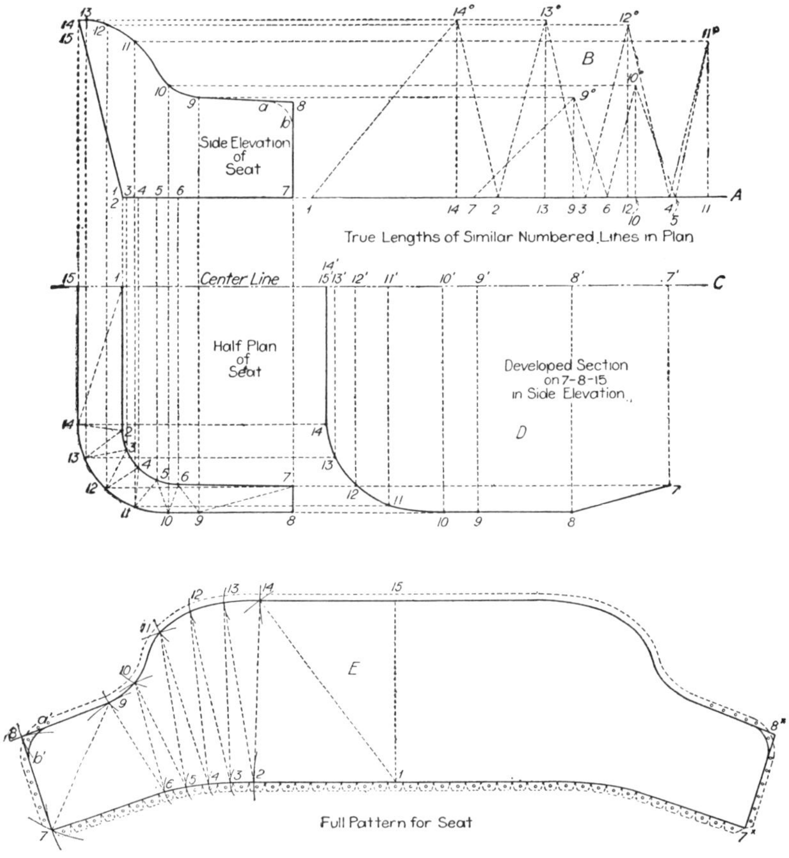

Fig. 67. Developed Section, True Lengths and Full Pattern for Two Passenger Seat Shown in Fig. 66

position as shown a half plan of the seat indicated by 1 7 8 15. In plan, 1 4 7 then represents the horizontal section through the seat 2 7 in elevation, while 8 12 15 in plan shows the horizontal view on the upper line of the seat 8 12 15 in elevation. By referring to Fig. 66, it will be noticed that the front corner of the seat at B is rounded, while a similar corner in the side elevation in Fig. 67 is square as shown at 8. This corner is made square so as to save time in developing the pattern, because the desired rounded corner indicated by the dotted curve *a b* in the side elevation can be drawn at pleasure in the pattern E, as there indicated by *a′ b′*, as will be explained in sequence.

The side elevation and half plan having been drawn in their proper position, the curved parts in the plan are divided into equal spaces as follows: Divide the inner curve of the plan into an equal number of spaces as shown from 2 to 6. In a similar manner divide the outer curve in the plan into a similar number of spaces as shown from 10 to 14. From the point where the curve meets the straight line in elevation at 9 drop a vertical line intersecting the plan view also at 9. Number the points in the plan as shown from 1 to 15 and connect them by dotted lines as shown. From the intersections 1 to 7 in the plan, erect vertical lines intersecting the seat line in the elevation also from 1 to 7. From points 8 to 15 in the plan, again erect vertical lines cutting the upper line of the seat in elevation, also from 8 to 15. These lines in the plan, then represent the bases of triangles which will be constructed, the altitudes or heights of which are equal to the vertical distances between similar numbers in the side elevation, and are obtained as follows:

Extend the seat line 2 7 in elevation as 2 A, and on this line place the various distances in plan as from 1 to 14, 14 to 2, 2 to 13 etc; up to 9 to 7, as shown by similar numbers on the line 2 A. All points numbered 1 to 7 remain on this line, as they represent the points in the seat line in the side elevation. All points from 9 to 14 on the line 2 A are projected vertically as shown, and are intersected by lines drawn parallel to 2 A from similar numbers in the side elevation, thus locating the altitudes of the triangles from 9° to 14°. Draw slant lines from 1 to 14°, 14° to 2, 2 to 13°, etc., until the last line 9° to 7 is drawn. These slant lines in B then represent the true lengths of similar numbered lines shown in the half plan view.

The next step is to find the developed section or true edge line along the line 7 8 15 in the side elevation, which is obtained as follows: Extend the center line in plan as 15 C. Take the girth of the top and front of the seat in the side elevation as shown from 15 to 8 to 7 and place these various spaces as shown by similar numbers 15′ to 7′ on the line 15 C. At right angles to 15 C and from the small figures draw lines, which intersect by lines drawn parallel to 15 C from similar numbered points 15 to 7 in the plan, thus obtaining similar intersecting points 7 to $15'$ in the developed section D. A line traced through points thus obtained as shown will be the developed section on 7 8 15 in the side elevation.

The pattern is now in order. As 1 15 in the side elevation shows its true length, take this distance and place it on any vertical line as 1 15 in E. With $15'$ 14 in the developed section D as radius and 15 in the pattern E as center describe the arc 14, which intersect by an arc struck from 1 as center and 1 14° in the true lengths in B as radius. With 1 2 in the half plan as radius and 1 in E as center describe the arc 2, which intersect by an arc struck from 14 as center

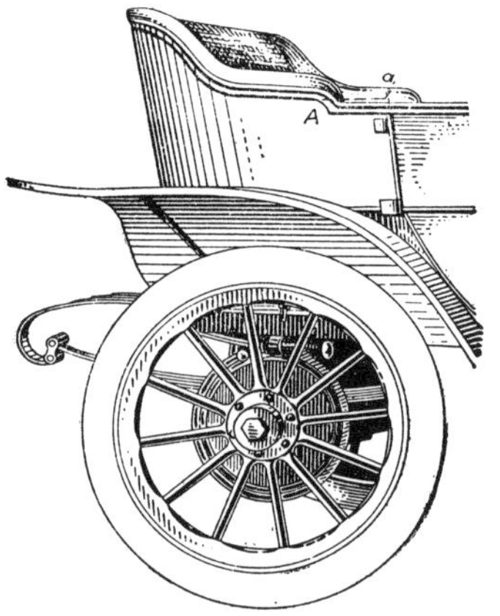

Fig. 68. Seat with Different Outline

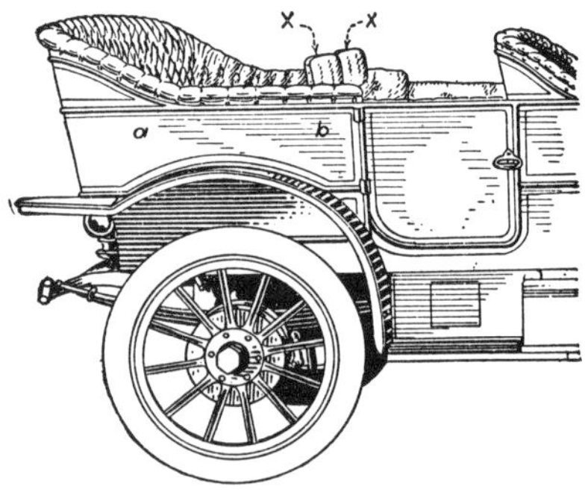

Fig. 69. Outline of Deep Seat in Touring Car

and 14° 2 in the true lengths in B as radius. Proceed in this manner, using alternately first the proper division in D, then the true length in B, the proper division in the inner half plan, then the true length in B until the line 10 6 in the pattern E has been obtained. Then using 10 9 in D as radius and 10 in E as center, describe the arc 9, which intersect by an arc struck from 6 as center and 6 9° in B as radius. With 6 7 in plan as radius and 6 in pattern E as center, describe the arc 7, which intersect by an arc, struck from 9 as center and 9° 7 in B as radius. With 9 8 in D as radius and 9 in E as center, describe the arc 8, which intersect by an arc struck from 7 as center and 7 8 in D as radius. Trace a line through points thus obtained in E. Then will 1 7 8 15 be the half pattern, which is traced opposite the line 1 15 as shown by 7^x 8^x. The full pattern is thus represented by 7 7^x 8^x 8 for a two passenger seat.

At 8 and 8^x in the pattern the desired shape of rounded corner can be drawn as indicated at the left at a' b'. Laps and flanges are allowed as shown by the dotted lines. Those at the bottom are notched and punched and nailed to the seat frame, while the lap along the top is drawn inward with hammer and dolly as will be explained later.

When the two passenger seats have different curvatures at the arm of the seat, as *a b* in the side elevation, it is well in developing the pattern to assume that the arm will have a corner as at 8 and then cut off as much from the pattern E as there indicated by *a′ b′*. Thus, if the arm of the seat were to have a curve as shown at A in Fig. 68 we would develop the pattern as if the arm were to meet at *a*, then trim off the pattern for the desired curve at A. Again, if the pattern were required for a deep seat as shown in Fig. 69 in front of which the two seats X X are placed, all that would be required would be to add the length of the arm *a b* to the pattern E in Fig. 67 from 9 to 8.

Constructing the Seat

When the pattern for the seat is developed, it is cut from No. 26 gauge sheet iron to be used as a template when cutting the desired number from sheet metal. This pattern E is cut out with the hand shears, but the notches at the bottom as well as the holes in the flanges are cut on a small power press, which punches four notches at a time and also punches the holes in the notches at the same operation ready to nail to the seat frame.

After the seats have been cut they are passed through the rollers to take out any buckles or dents and to smooth same, after which the bead is put in the seat pattern, the proper size beader (usually ½ in.) being used on a swedging machine. The bead is made flush at the bottom and 3-16 in. away from the top and front edges as shown in the part pattern in Fig. 70. These beads are put in on the flat

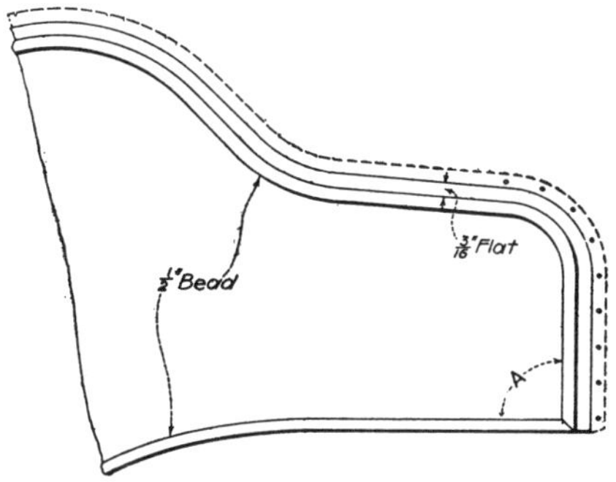

Fig. 70. Part of Pattern for Seat Showing Bead around Same

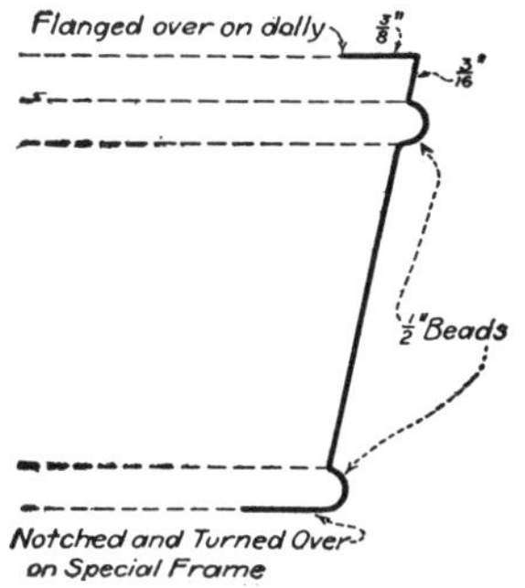

Fig. 71. Sectional View through *a b* in Fig. 66

sheets and formed or shaped afterward by hand. If these seats were stamped up in one piece this beading on the flat would not be necessary as the power press would do this work in one operation. But as a rule the seats are usually beaded and shaped by hand. When beading these seats on the swedger it is impossible to form the miter at the corner A, which must be finished by using steel die plates and a special hammer, as was explained in connection with Figs. 11 and 12.

Fig. 71 shows a sectional view through *a b* in Fig. 66, and makes clear how the flanges and beads are placed in the seat body. After the seats have been beaded in the flat, they are formed to the required shape over a tapering mandrel stake, after which they are shaped true over a wooden template and will look as shown by A in Fig. 72, one template being used for the top and another for the bottom.

When the shaping has been done the bending of the bottom notched flanges are in order. These flanges could be turned over with flat plyers, but this method would take too much time and would not make as sharp a bend as can be made on the angle iron template shown in Fig. 73. This template is made of heavy angle iron 2 × 2 × 3/8 in. having the required shape to fit inside of the bottom of the seat. Holes are punched through the frame at *a* and *b*, which allows the frame to be bolted to a special small bench made for that purpose, as shown by *a* in Fig. 74.

After the seats have been shaped up true to the template they are set over the angle iron frame, as shown by *b*, when the notched flanges may be turned over

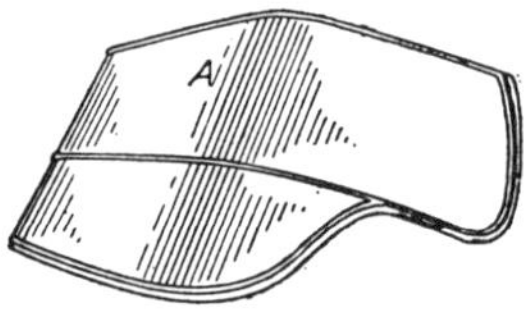

Fig. 72. Shaping the Seats Over Wooden Template

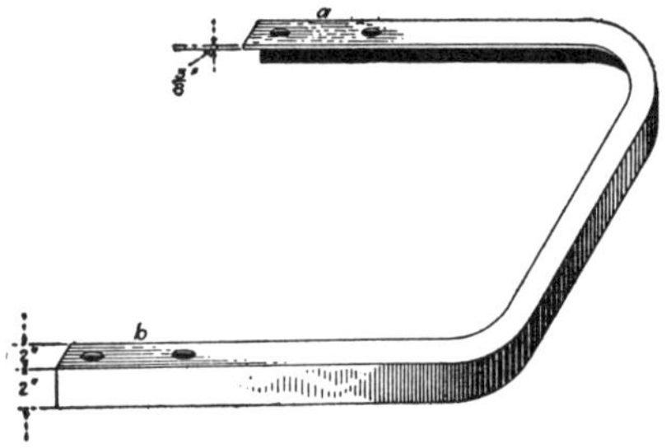

Fig. 73. Angle Iron Template for Bending Over the Lower Flanges of the Seat

with the mallet, and when completed will look as shown in the illustration in Fig. 75. The metal is now nailed to the ash seat with 1 in. No. 14 gauge wire nails, after which it is clamped to the wooden horses by means of the clamps *a* and *b* in Fig. 76, ready for setting down the upper flange of the metal seat. Before this flange is turned over, an iron frame having the proper shape and shown by *c* is clamped to the upper part of the seat, as shown by *d*, after which the flange can be turned over, by using the dolly *f* in one hand and the hammer *e* in the other, as shown. By having the iron frame or template clamped in the position shown, this helps to retain the shape of the seat at the top after the flange is turned in and the template *c* is removed. The detail of the dolly used for this work is shown in Fig. 77 by A, made so as to fit in the bead, *a* being the desired width or 3-16 in.

In dressing down the flange over the dolly A, light blows should be struck with the iron hammer B and afterwards dressed smooth with the mallet.

The seats are now taken in the carpenter shop and the front part of the side of the seat lined with ash framing as shown by A in Fig. 78, and the upper flange of the seat nailed to the wood work. Malleable iron angles are then bolted to the seat frame and through the metal with $\frac{1}{4} \times 1\frac{1}{2}$ in. stop bolts. To secure rigidity to the metal seat, iron seat braces are bolted to the frame *a b* and to the metal back

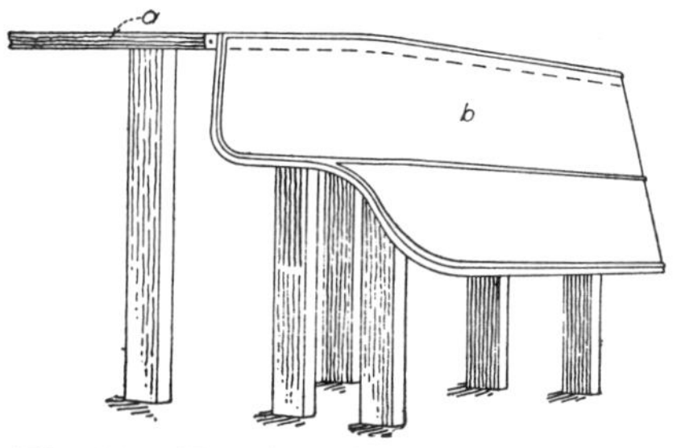

Fig. 74. Turning Over the Notched Flanges of the Seat

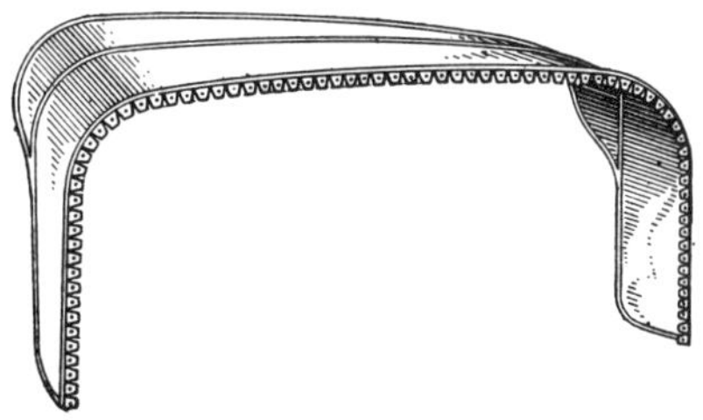

Fig. 75. Appearance of Flanges When Turned Over

Fig. 76. Turning Over the Upper Flange with Hammer and Dolly

at D E. These braces are usually $1\frac{3}{4}$ in. wide and $\frac{1}{2}$ in. thick, tapering to $\frac{1}{4}$ in. at the ends. When this has been done the seat is fitted to the sheet metal body of the car, as shown by A and B in Fig. 79, $\frac{1}{4} \times 1\frac{1}{2}$ in. stop bolts being used to bolt down to the body as well as to the panels between the sides of the body. There is also a brace known as a T iron, which is bolted to the body of the car below the rear seat on each side as shown by K in the scale drawing in Fig. 2, and connects and is bolted to the side brace in the seat at L, with $\frac{1}{4} \times 1\frac{1}{2}$ in. stop bolts, thereby giving rigidity to the entire body and seat.

The front seat shown in Fig. 79 by C is known as an individual seat. While in a two passenger seat just described there is one straight-back, an individual seat has separate curved backs joined together at the center as shown, on which the rear seat brace is bolted. After the seats have been fitted, they are numbered to correspond to the number on the body, in this case the number being 482, representing the 482d car constructed of this type. The rear or two passenger seat is now loosened and set upon two wooden horses as in Fig. 78, and the upper ash frame B F C bolted through the metal flange of the seat, by means of 3-16 × 1 in. stove bolts with countersunk heads as indicated by *d*. The same applies to the individual seat which is shown on the carpenter's bench in Fig. 80. Note the side seat brace at D, the rear seat brace at A B C, and the angles to which the outside coat rail is riveted at E and F. Over the rear seat brace at B C the ash frame is bolted, as well as along the top as shown, ready to receive the nailing of the upholstery work.

After all of the sheet metal parts of the body have been accurately fitted, they are sent to the sand blasting room as shown in Fig. 81. This is a room lined with

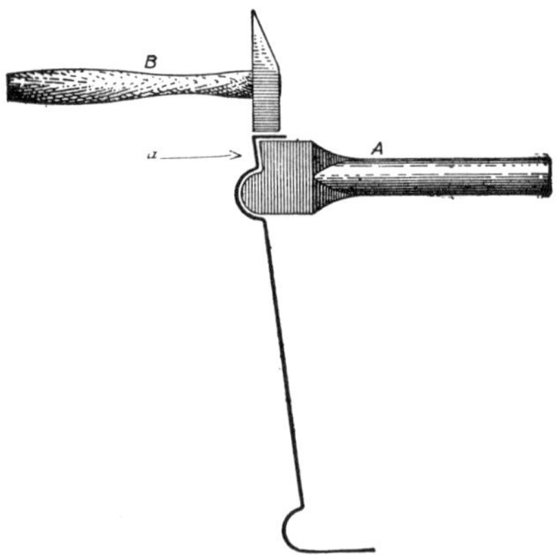

Fig. 77. Shape and Use of the Dolly in Flanging Top of Seat

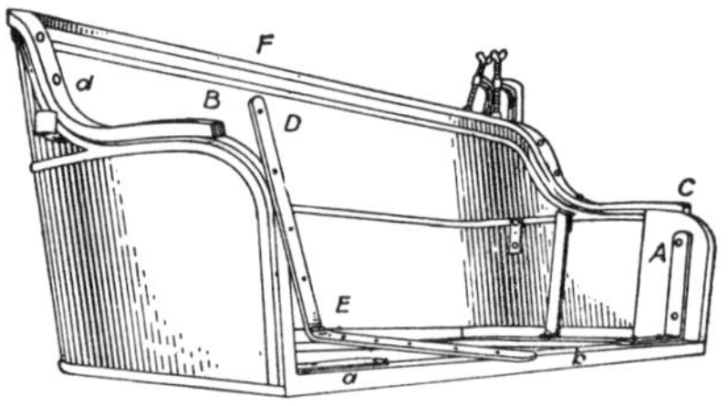

Fig. 78. Bolting Ash Frames on Metal Seat

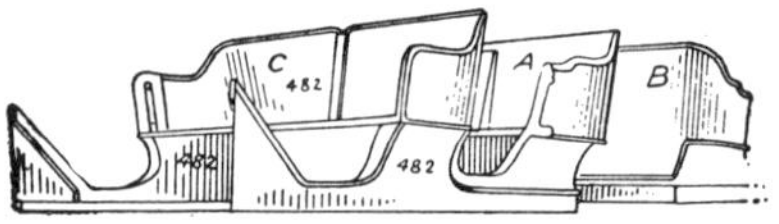

Fig. 79. Fitting the Seat to Body of Automobile

corrugated iron in which the surfaces of the metal going into the manufacture of the automobile are cleaned from rust, scale, and dirt, by means of the sand blast through the nozzle A, giving an ideal surface, on which a lasting finish can be produced in enamel. After the seats have been enameled they are upholstered and then bolted to the enameled body, provision being made to allow the bolts to pass through the seat and body without disturbing the upholstering.

Pattern for Individual Seats

The next type of seat to be developed will be that of the individual seat, a finished view of which is shown in Fig. 82. This seat as shown in the cut, contains two separate seats and is developed as shown in Fig. 83 in which 5 11 12 19 shows the side elevation of the seat, while 1 2 5 6 11 12 18 19 22 23 in the half plan, shows one half of the double seat below the center line A B. The side elevation in its proper relation to the plan, having been drawn, the first step is to divide the curves, 2 to 5 and 22 to 19 each into equal spaces, in this case three as shown. In a similar manner divide the quadrants 6 to 10 and 14 to 18 into four parts each as shown. From 14 the end of the quadrant in plan, erect a vertical line into the elevation cutting the top line of the seat at 14, which, however, is not the end of the curve in the top line of the seat. Where this curve meets the straight line as at 13^v, drop a vertical line intersecting the top line of the seat in plan at 13. Now connect the points in plan as shown from 22 to 3, 3 to 21, 21 to 4, etc., until the last line 13 to 11 has been drawn. From the various points 1 to 11 in the bottom of the seat in plan project lines to the elevation, cutting the bottom of the seat as shown by similar numbers 1 to 11. In a similar manner from the various points 12 to 23 in the top of the seat in plan project vertical lines into the elevation, cutting the top line as shown from 12 to 23. The lines in the plan will then represent the bases of triangles which will be constructed in G, the vertical heights or altitudes of which will be equal to the vertical heights between similar numbers in the side elevation. The center line shown by 1 23 in plan, is shown in its true length by the dotted line 1 23 in the side elevation. To obtain the rest of the true lengths of the

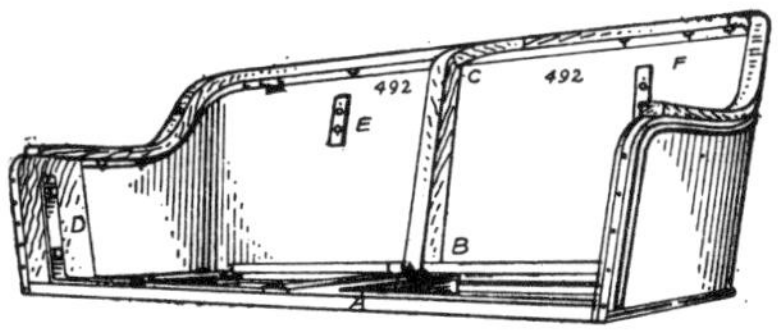

Fig. 80. Completing the Ash Framing on an Individual Seat

Fig. 81. Sand Blast Used in Cleaning the Metal Work Used on Automobiles

dotted lines shown in plan, take all the separate distances of the various dotted lines and place them as shown by similar numbers in G on the line D E. At right angles to D E and from points 13 to 22 inclusive, erect vertical lines, which intersect by horizontal lines drawn from similar numbers in the upper part of the seat in the side elevation, thus obtaining intersections 13° to 22°. From these intersections draw slant lines to the proper numbers, as called for in the plan view. These slant lines in G will then represent the true lengths of similar numbered lines in the half plan.

The true edge line on the line 11 12 19 in the side elevation must now be obtained. Extend the line A B in the plan as B C, upon which place the girth of the various spaces in 19 12 11 in the side elevation, as shown by similar numbered points 19′ to 11′ on B C. At right angles to B C from these small figures draw lines, which intersect by lines drawn parallel to B C from similar numbered points in the half plan. Trace a line through points thus obtained in F, as shown from 11 to 12 to 22 to 23′, which will be the developed section or true edge line desired.

The pattern may now be laid out as follows: Take the length of 1 23 in the side elevation, which shows the true length through the center, and place it on the

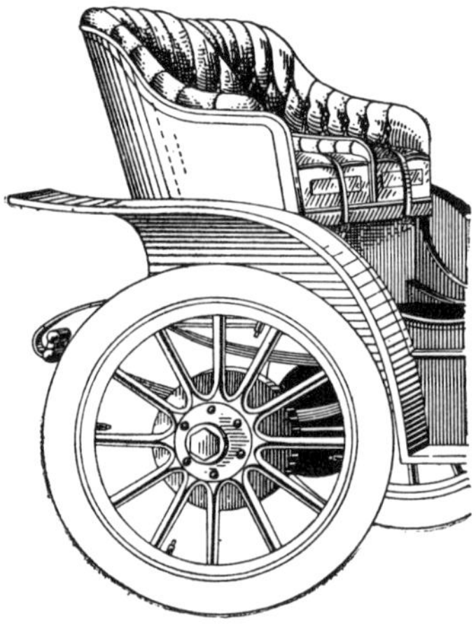

Fig. 82. Finished View of Individual Seats

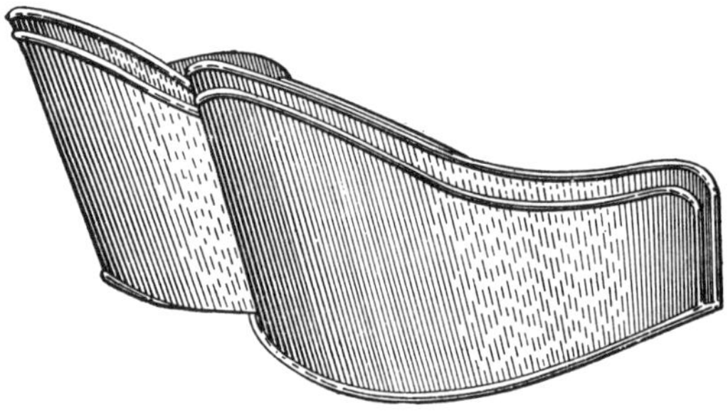

Fig. 84. Rear View of Beaded Individual Seats without Wooden Frames

vertical line in H as shown by 1 23. At right angles to 1 23 draw the lines 1 2 and 23 22 equal to similar numbers in the plan view, and draw the line in H from 2 to 22. Now with 2 3 in the half plan as radius and 2 in H as center, draw the arc 3, which intersect by an arc struck from 22 as center and 22° 3 in the true lengths in G as radius. With radius equal to 22 21 in the developed section F and 22 in H as center, describe the arc 21, which intersect by an arc struck from 3 as center and 3 21° in G as radius. Proceed in this manner in developing the rest of the pattern. The spaces from 21 to 11 in the pattern H are obtained from similar numbered spaces in F, while the spaces from 3 to 11 in H are obtained from similar numbers in the half plan. The length of the dotted lines in the pattern H are obtained from similar numbered slant lines in the true lengths in G. A line traced through points thus obtained in H, as shown from 1 to 2 to 11 to 12 to 22 to 23 will be the half pattern. Trace this half opposite the line 23 1 as shown by similar numbers to the right. Then will 11 1 11^{v} 12^{v} 23 12 11 be the full pattern for the individual seat, to which flanges should be

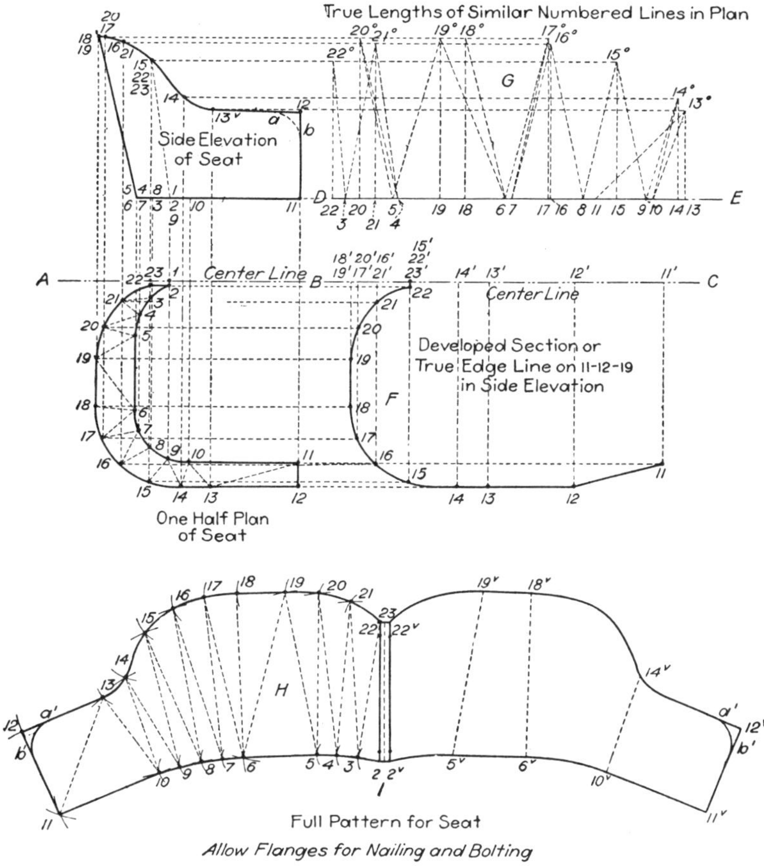

Fig. 83 Side Elevation, One Half Plan, Developed Section, True Lengths, and Full Pattern for Individual Seat Shown in Fig. 82

allowed for nailing and bolting to the wood frames. The curves a' b' are added to each corner as required.

When this seat is made up and beaded, it will look as shown in Fig. 84. After the seat has been fitted to the body, the ash frame work is put in place as previously shown in Fig. 80.

Pattern for Single Rumble Seat

The next type of seat to be developed is shown in Fig. 85 and is known as a single rumble seat. In this case the seat may be developed by using the cone method or by the use of radial lines as is explained in detail in Fig. 86, in which A B $2'$ $2''$ shows the side elevation of the seat, above which is drawn the half plan in its proper position as shown, the center for striking the rounded corner being indicated by d. The seat in this case has a similar flare at the back and two sides as shown in plan by x and y. The inner line in the half plan shows a section through the bottom of the seat 2″ A in the side elevation, while the outer line in the half plan shows the true section on the horizontal line 2′ f in the side elevation.

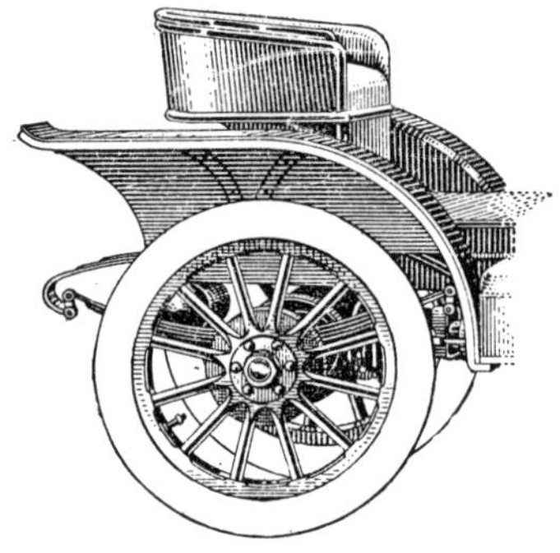

Fig. 85. Finished View of a Single Rumble Seat, Developed by Radial Lines

The pattern for the rounded corner will be first developed, and to it will be added the back and sides of the seat. First divide the outer quadrant 2 5 in plan into equal spaces as shown by the small figures 2, 3, 4 and 5, from which points drop perpendicular lines meeting the horizontal line 2′ f in the side elevation at 2′, 3′, 4′ and 5′. From the center point d in the plan draw the vertical line d e, which intersect by extending the side of the seat 2′ 2″ in the side elevation at e. From the intersections 3′ and 4′ on the line 2′ f draw radial lines to e, cutting the upper line of the seat 2′ B as shown by the dots on 2′ B. At right angles to the center line of the cone 5′ e, from the various intersections on the top line of the seat 2′ B draw lines, which intersect the rear line of the seat 2′ 2″ as shown by the dots. Now using e as center and e 2′ as radius draw the arc 2′ 2. On this arc, starting from 2, lay off the girth of the quadrant 2 5 in plan as shown by similar numbers on the arc 2 2′. From these points 2, 3, 4 and 5, draw radial lines to e, which intersect by arcs drawn from e as center with radii equal to the various dots on the line 2′ 2″ in the side elevation. Through the points of intersections thus obtained draw the arc C D. Now using e as center and e 2″ as radius draw the arc E F as shown. The pattern for the rounded corner is then shown by C D E F.

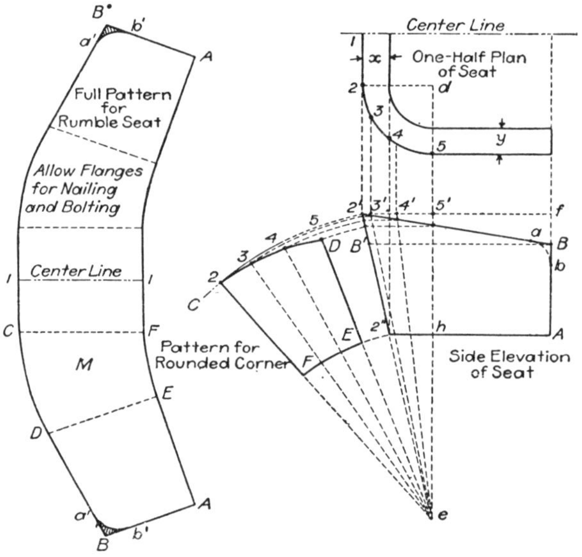

Fig. 86. Side Elevation, One Half Plan, and Pattern for Single Rumble Seat Shown in Fig. 85

To complete the half pattern for the seat proceed as follows: Take a tracing of C D E F and place it as shown by C D E F in M. At right angles to C F draw the lines C 1 and F 1 equal to 2 1 in the half plan. Draw the center line

1 1 in M. Now at right angles to D E draw the line E A equal to *h* A in the side elevation, and at right angles to E A in M draw the line A B, equal to the true length 2″ B′ in the side elevation, this true length being obtained by drawing a line from B parallel to 2″ A until it intersects at B^1. Having established B in M, draw a line from D to B and add the desired curve *a′ b′* to correspond with *a b* in the side elevation. The half pattern is then shown by 1 B A 1, which is transferred opposite the center line 1 1 as shown. The full pattern for the single rumble seat in Fig. 85 is shown by A A° B° B in Fig. 86.

Pattern for Single Rumble Seat, without Flare

Some rumble seats are made without flare, as shown in Fig. 87, when the pattern may be developed by parallel lines as shown in Fig. 88. This is a very simple method and will be briefly described. Let A B C D represent the side elevation of the seat, below which in its proper position is shown the half plan. This half plan is divided into equal spaces as shown from 1 to 6 and from these points lines are erected, cutting the outline of the seat in the side elevation as shown. Extend the line C D in the side elevation, as D E, on which place the girth of twice the number of spaces contained in the half plan, as shown from 1 to 6 to 1^v on D E. Through these small figures and at right angles to D E, draw lines, which intersect by lines drawn parallel to D E from similar intersections on A B in the side elevation. Trace a line through points thus obtained.

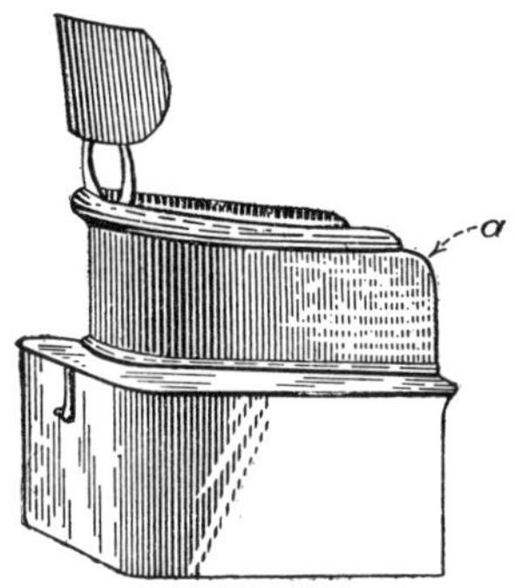

Fig. 87. Finished View of a Single Rumble Seat, Developed by Parallel Line

Then will 1^v F G H 1 be the full pattern shape, to which flanges must be allowed for nailing and bolting. At the corners of the pattern at H and F, draw the curves *a′ b′* to correspond to *a b* in the side elevation, and as shown in the finished view in Fig. 87 by *a*.

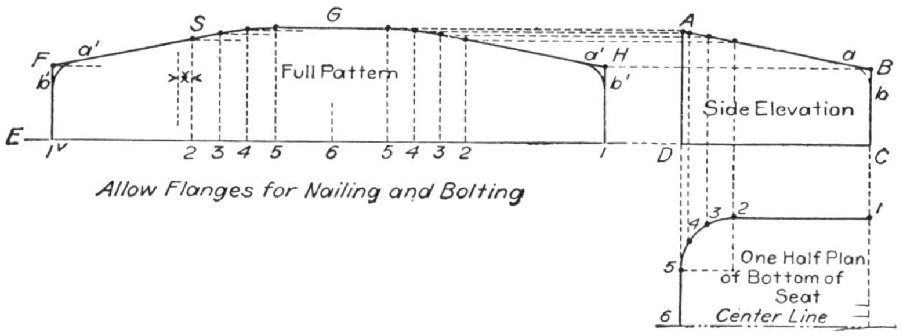

Fig. 88. Half Plan, Side Elevation, and Pattern for Single Rumble Seat as Shown in Fig. 87

Pattern for a Double Rumble Seat

Fig. 89 gives a finished view of a double rumble seat, the two separate seats joining together on the two inner curved corners. In this case the seat has a flare as shown and the two outer curved corners are developed by radial lines as explained in connection with Fig. 86. The inner corners, or the two rounded corners joining at the center, are so constructed that the parallel line method of development may be employed as shown in Fig. 90. So that the same pattern developed for the

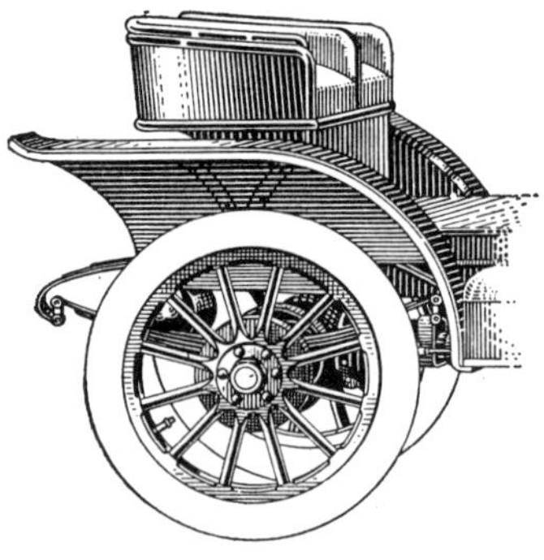

Fig. 89. Finished View of a Double Rumble Seat

curved outer corner in Fig. 86, can be used, the same size plan and elevation has been drawn in Fig. 90, the outer corner *d* in the half plan, being similar to the corner *d* in Fig. 86, the pattern for which is shown by C D E F, which can be used in laying out the completed pattern in Fig. 90.

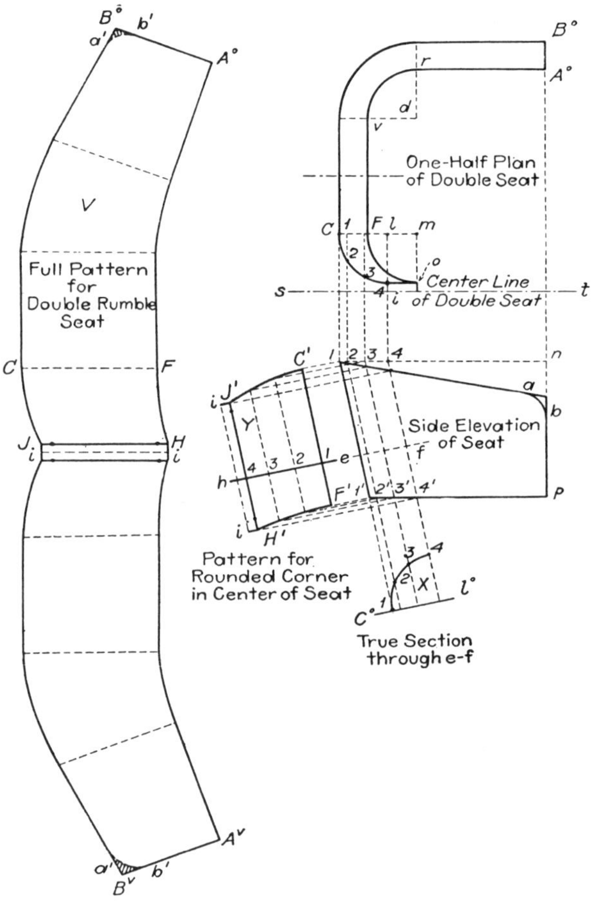

Fig. 90. Half Plan, Side Elevation, and Full Pattern for Double Rumble Seat Shown in Fig. 89

The method of drawing the plan of the rounded corner where it meets the center line *s t* is as follows: Using *d v* as a radius and *m* in the half plan as a center, draw the quarter circle F *o* making it a duplicate of *v r*. From the center *m* draw a line parallel to *t s* cutting the upper line of the seat in plan at C. Using C as a center and *m* F as a radius, describe an arc, cutting the line C *m* at *l*. Using the same radius and *l* as a center draw the quarter circle C 4. Draw a line from 4 to *o* and make 4 *i* equal to the half width of the flat space for the seat brace between the two inner corners.

The true section on the line 1 *n* in elevation is shown by B° C 4 *i* and A° *r v* F *o* the true section on the seat line 1′ *p* in elevation.

As C 4 *i* in the half plan represents a true section on the horizontal plane 1 *n* in elevation another true section must be found at right angles to the flare of the seat 1 1′ or on the line *e f* in elevation. First divide the quadrant C 4 in plan into any desired number of equal spaces, as shown by 1, 2, 3 and 4. From these points and at right angles to *s t* drop lines cutting the horizontal line 1 *n* in elevation, also at 1, 2, 3 and 4. From these small figures draw lines parallel to 1 1′ indefinitely cutting the seat line 1′ *p* in elevation at 1′, 2′, 3′ and 4′. At right angles to these lines, draw any line as C° *l*° as shown. Now, measuring from the line C *l* in the half plan, take the various projections to points 2, 3 and 4 and place them on similar numbered lines, measuring in each instance from the line C° *l*°, thus obtaining the points 2, 3 and 4 in X. Trace a line through points thus obtained and then will X be the true section through the line *e f* in the side elevation, from which the true girth may be obtained in developing the pattern.

Extend the line *f e* as *f h*, on which place the girth of the true section X, as shown by similar numbers on *e h*. Through these small figures and at right angles to *h e* draw lines as shown, which intersect by lines drawn parallel to *h e* from similar points on the upper seat line 1 *a* and the lower seat line 1′ *p*. Trace a line through points thus obtained as shown by J^1 C^1 F^1 H^1, which will be the pattern for the rounded corner in the center of the seat. At right angles to J^1 H^1 draw J^1 *i* and H^1 *i* equal to 4 *i* in the half plan and draw a line from *i* to *i* in Y.

By using this pattern in connection with part of the one shown in Fig. 86, the full pattern for the double rumble seat may be drawn as shown by V in Fig. 90. Of course it should be understood that these two patterns in Figs. 86 and 90 can only be used because both sizes and shapes of plans and elevations are similar. That part of the pattern shown in Fig. 86 by A° B° C F A° is the pattern, for that part of the seat shown in plan in Fig. 90 by A° B° C F A° and is reproduced as shown by A° B° C F A° in V. On the line C F in V place a reproduction of C^1 F^1 H^1 *i i* J^1 in Y as shown by similar letters in V. Then A° B° *i i* A° is the half pattern of the double rumble seat. Trace this half opposite the line *i i* as shown by A^v B^v. Then A° B° B^v A^v is the completed pattern for the double rumble seat, to which flanges for nailing and bolting must be allowed. Cut off the rounded corners shown shaded by *a′ b′* in V. If a double rumble seat were desired where the seat has no flare and stands vertical as in Fig. 87, then all that would be necessary after the pattern for the single seat in Fig. 88 is developed, would be to simply turn over the part pattern S H 1, 2 opposite the line S, 2, allowing a flat surface for the seat brace as shown at *x*.

Pattern for a Racer Seat

Fig. 91 shows an odd shaped seat in a racer car, requiring triangulation in its development. Note that the seat bottom pitches towards the back, thus giving a comfortable position to the rider when speeding. The method of laying out this pattern is explained in detail in connection with Fig. 92, in which 1 6 7 15 shows the side elevation of the seat and 1 2 6 7 12 15 the half plan of the racer seat, the center line through the seat being indicated. As the corner curve on the top line of the seat, is considerably larger than the curve of the bottom of the seat as shown in plan, the method of placing the divisions in these curves is as follows: Divide the quadrant in the bottom of the seat in plan as shown by 2, 3, 4 and 5 or in this case three spaces. In a similar manner divide the quadrant in the outline of the top in plan as shown by 10, 11, 12, 13 and 14 or in this case four spaces. From 10 in plan erect a vertical line to the elevation, cutting the top outline of the seat also at 10. Now divide the curve between 10 and 8 in elevation into two or any desired number of spaces, as shown by 9 and 8 from which points drop lines cutting the outline in plan also at 9 and 8. Number all the points in the half plan, as shown from 1 to 15 and connect the unequal number of spaces in the quarter circles in a manner as indicated by the dotted lines. From the various spaces 1 to 6, lines are erected cutting the seat line in elevation also from 1 to 6. In a similar manner from the various spaces 7 to 15 in plan erect lines cutting the top line of the seat in elevation also from 7 to 15. The various lines in plan then represent the bases of triangles which will be constructed in B, the altitudes of which will equal the various perpendicular heights between similar numbers in the side elevation.

Fig. 91. Finished View of a Racer Seat Requiring Triangulation in Its Development

Note that the seat is to have a curved corner in the side elevation as shown from *a* to *b*. This curve will be drawn free hand in the pattern, and for this reason are the lines 6 *b* and 8 *a* extended until they meet at 7.

When the various true lengths of the base lines in plan are being found it will be noticed that these base lines are set off on similar numbered horizontal lines, drawn from similar numbers in the seat line in elevation, and then lines connected to similar numbered points obtained from the top line of the seat in the side elevation. To make this clear, it will be shown as an example how the true length of the line 5 10 in plan is found. From 5 in the side elevation draw a horizontal line as shown, on which place the distance 5 to 10 in plan as shown from 5 to 10 in diagram B. From 10 erect the perpendicular line 10 10°, intersecting it with a horizontal line drawn from the point 10 in the side elevation, at 10°. Draw a line from 10° to 5 which gives the true length of the line 10 5 in plan. In this manner all of the true lengths are found. Thus the true length of 8 6 in plan is found by setting off this length on the horizontal line drawn from 6 in the side elevation as from 8 to 6 in B, and from 8 a perpendicular line erected and intersected by a horizontal line drawn from 8 in the side elevation at 8°. The distance from 8° to 6 then shows the true length of 8 6 in the plan.

The next step is to find the true edge line of the top, front, and bottom of the seat as shown in A, which is obtained as follows: Take the girth of the various spaces contained in the side elevation from 1 to 6 to 7 to 15 and place each one separately on the center line extended in the plan, as shown by similar numbers from 1′ to 15′. From these points and at right angles to the center line draw lines as shown, which intersect by lines drawn parallel to the center line from similar numbered points in the half plan. Trace a line through points thus obtained, as shown from 1′ to 5 to 12 to 15′ which will be the true section or edge line, on similar numbered outline in the side elevation.

The pattern shape is now in order and is developed as follows: As 1 15 in the side elevation shows the true length on the line 1 15 in plan, take this distance and place it on the vertical line in C as shown by 1 15, With 15′ 14 in the true section A as radius and 15 in the pattern C as center, describe the arc 14, which intersect by an arc struck from 1 as center, and 14° 1 in the true lengths B as radius. Then with 1′ 2 in A as radius and 1 in C as center, describe the arc 2, which intersect by an arc struck from 14 as center and 14° 2 in B as radius. Proceed in this manner until the pattern is developed. The various divisions from 1 to 6 to 7 to 15 in the pattern C are obtained from similar numbered divisions in A, while the true lengths of the dotted lines in C are obtained from similar slant lengths in diagram B. In connecting the various divisions in C the connecting numbers in the plan are used as a guide; in other words, whichever way the numbers are connected in plan in the same manner will the numbers be connected in the pattern. A line traced through these points as shown from 1 to 6 to 7 to 15 in C will be the half pattern. Trace this half opposite the line 1 15, as shown by 6′ 7′. Then will 6 7 15 7′ 6′ 6 be the full pattern, to which flanges are allowed for nailing and bolting.

Referring to the side elevation it will be noticed that the corner is rounded as at *a b*. The method of projecting these points accurately on to the pattern is as follows: Take the distances from 8 to *a* and 7 to *b* and place them on the center line in plan, as shown from 8′ to *a*′ and 7′ to *b*′. From *a*′ and *b*′ draw perpendicular lines cutting the true edge line at *a* and *b* respectively. Now with radius equal to 7 *a* and 7 *b* in A and using 7 in C as center, describe the arcs *a* and *b* respectively. Draw the desired curve from *a* to *b* as shown.

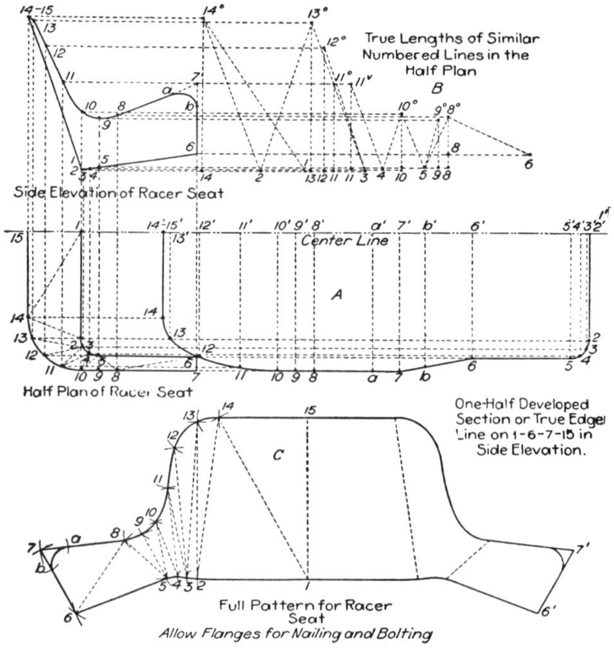

Fig. 92. Side Elevation, One Half Plan, Developed Section, True Lengths, and Full Pattern for Racer Seat Shown in Fig. 91

Should an individual seat be desired of this type of racer seat, the same method and principle should be employed as was explained in connection with the individual seat shown in Fig. 83, in which case the outer corners of the seat remain the same, but the middle corners where the two seats meet, change in shape as shown. It makes no difference what the shape of the corners may be the principles are similar in either case.

Pattern for a Surrey Seat

In Fig. 93 is shown a view of a finished surrey seat A, the back of which is not finished with upholstering, the top and front edge of the seat being beaded. As the back and sides of the seat are straight the pattern may be developed by parallel lines as indicated in Fig. 94, in which the side elevation of the seat as well as the half plan is shown. The pattern for a seat of this kind is easily obtained as follows: Divide the curved part in plan as shown and number all the points of intersections, starting at 1 and ending on the center line *a b* at 6. From these intersections perpendicular lines are erected cutting the outline in the side elevation also from 1′ to 6. Extend the bottom line of the seat in elevation as shown by B A, on which place twice the girth of the half plan, as shown from 1 to 6 to 1° on the line A B. From these divisions and at right angles to A B draw perpendicular lines intersecting them by lines drawn parallel to A B from similar

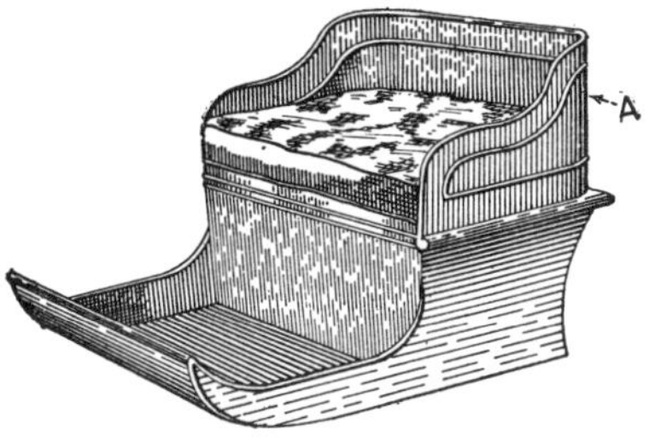

Fig. 93. Finished View of a Surrey Seat with Beaded Edges

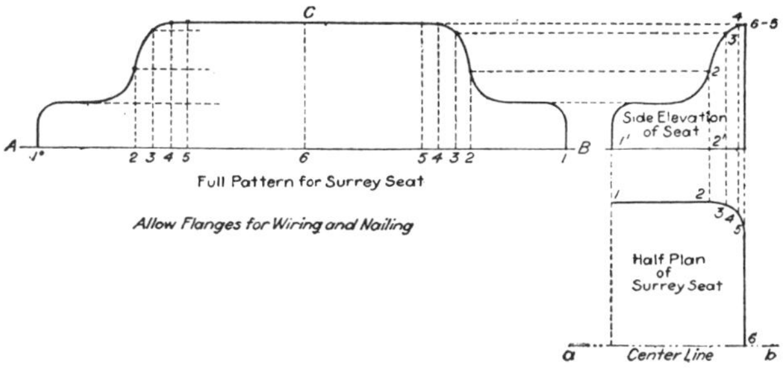

Fig. 94. Plan, Elevation, and Pattern for Surrey Seat Shown in Fig. 93

numbered intersections in the outline in the side elevation. A line traced through points thus obtained, will give the full pattern shape for the surrey seat shown in Fig. 93. As the top and front edge of the seat is wired, allowance must be made along A C B in the pattern in Fig. 94 for same, as well as flanges allowed along 1 1° for nailing to the wooden seat frame.

Pattern for Motor Truck Seat

Fig. 95 shows a view of a finished motor truck seat indicated by A, which is developed by means of parallel lines as shown in Fig. 96. In this figure A shows the side elevation of the seat and B C D the half plan. As the flare of the seat is the same on three sides then 1 2 in the side elevation gives the amount of material required in developing the pattern. At right angles to the lines of the seat in A draw the girth line, on which place the girth of 1 2 in the side elevation as shown by 1′ 2′, through which draw

Fig. 95. Finished View of a Motor Truck Seat.

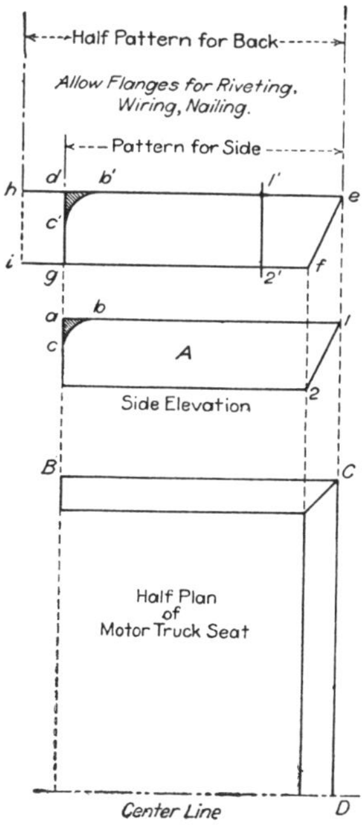

Fig. 96. Plan, Elevation, and Pattern for Motor Truck Seat, Shown in Fig. 95

the horizontal lines as shown, and intersect them, by lines drawn parallel to 1′ 2′ from similar numbered divisions in the side elevation. Trace a line through points thus obtained, as shown by *d e f g*, which will be the pattern for the sides of the seat. As the corner *a* in the side elevation is desired to be rounded as shown by *b c* simply reproduce a similar curve in the side of the pattern as shown by *b′ c′*. For the pattern for the back simply take the length from C to D in the half plan and set it off from *e* to *h* in the pattern and draw the perpendicular *h i* until it intersects the line *f g* extended at *i*. The half pattern for the back of the seat is then shown by *h e f i* which should be traced opposite the line *h i* to obtain the full pattern. In this as in the previous seats, flanges must be allowed for riveting the corners, wiring the top and front, and nailing to the seat frame.

Constructing the Sheet Metal Door

In laying out the pattern for the door shown by A in Fig. 97 no drawing or pattern drafting is necessary, as the shape is obtained from the ash frame prepared by the carpenter. So that the constructive features may be thoroughly understood,

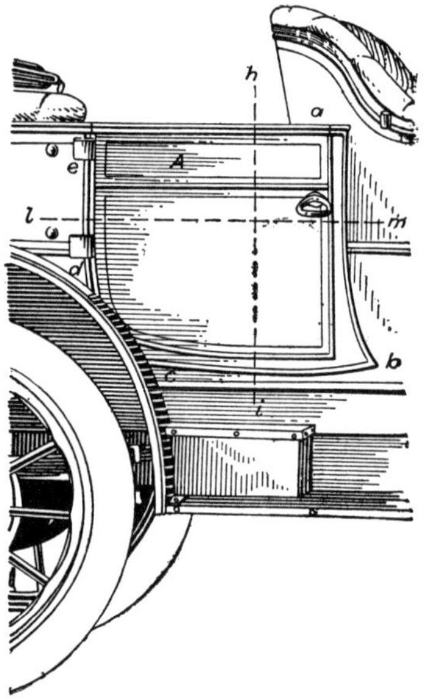

Fig. 97. View of Finished Door in Sheet Metal

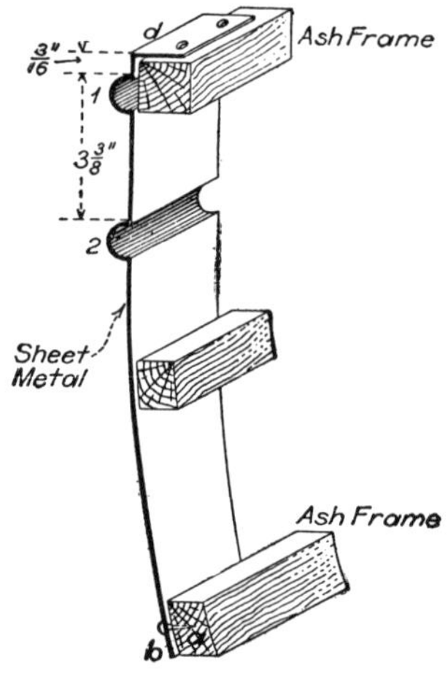

Fig. 99. Vertical Section through *h i* in Fig. 97

a detailed description of finishing the sheet metal door will be given. That part (the doorway frame) shown by *a b c d e* in Fig. 97 is made of cast aluminum against which the door closes. This cast aluminum molding is riveted to the metal body and seat with a neat finish so that no joint can be detected when the entire body is enameled. The method of fitting the ash frame of the door is shown in the view in Fig. 98. This shows the seats fitted to the body, on which the aluminum mouldings *a b c d* have been riveted and inside of which, the door frame *e* is being fitted. In fitting this ash frame *e* allowance is made along the sides and bottom to admit the thickness of the half round faced T iron, which is screwed over the metal face of the door, as will be described later.

After the wooden door frame has been correctly fitted, it is laid over a piece of No. 28 gauge black sheet iron and a template cut. No flanges are allowed to the

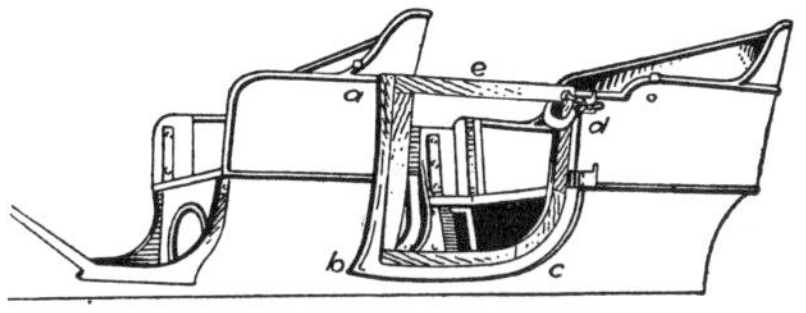

Fig. 98. Fitting the Wooden Door Frame to the Cast Aluminum Moulding Trimmings

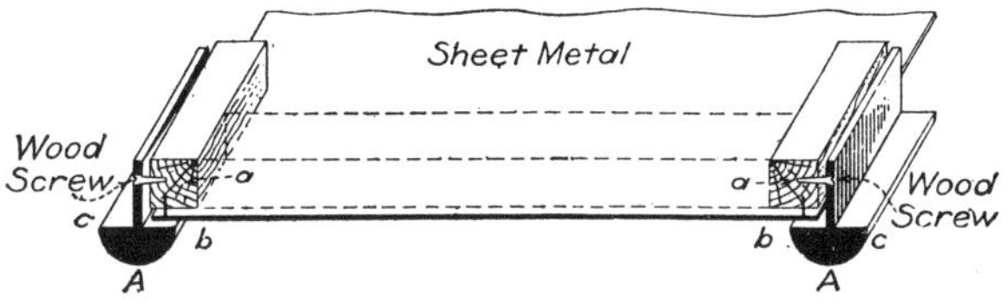

Fig. 100. Horizontal Section through *l m* in Fig. 97

template at the two sides and bottom, but a nailing flange is allowed along the top as shown in Fig. 99, at *d*, and allowance made for the two beads at 1 and 2. The metal doors are now cut from No. 20 gauge sheet steel or sheet aluminum, and cut on the power rotary shears as near to the outline as possible, and then if necessary trimmed with the hand shears. The door has a slight curve as shown from *b* to *d* in Fig. 99 and is obtained in the metal, by passing it through the rollers, after which the ½ in. beads 1 and 2 are obtained by passing the metal door through the power swedging machine, and the flange *d* is bent on the brake. Fig. 99 also shows a vertical section through *h i* in Fig. 97 and shows how the upper flange *d* is nailed to the top of the ash frame. At the bottom of the ash frame *a*, the metal is nailed flush at *b*, with 1-16 in. thick brads ¾ in. long. The distances between the swedges are shown full size or can be made as desired.

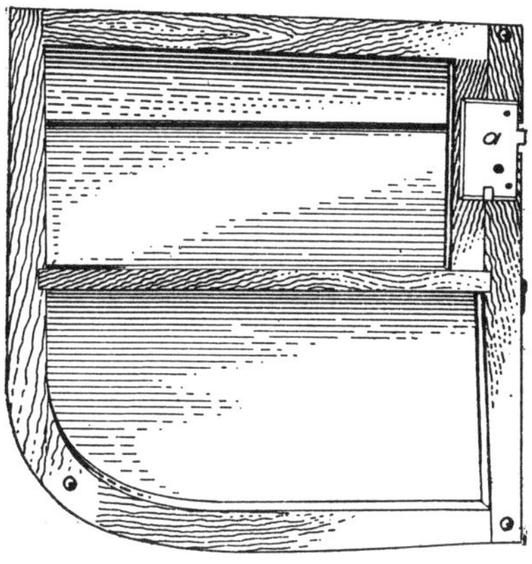

Fig. 101. Showing the Inside of the Wooden Door Frame, with the Metal Nailed on the Outer Side

Fig. 100 shows a horizontal section through *l m* in Fig. 97, and how the two side edges of the door are nailed flush with the wooden frame at *a* and *a*, in Fig. 100, and also the round faced T iron A A which is screwed to the sides and bottom of the ash frame with wood screws as indicated, the one side of the T iron at *b b* covering the brad heads *a a*, while the opposite angle at *c c* sets over the aluminum moldings *a c b d* in Fig. 98. Where these round faced T's intersect each other at the corners or join to the swedges in the sheet metal door, the iron T is filed, mitered, and soldered, and then scraped smooth. When this has been done, the door has the appearance shown by A in Fig. 97 in which the horizontal beads are formed in the sheet metal and the beads along the two sides and bottom are on the

Fig. 102. Two Other Styles of Doors

T irons. In Fig. 101 is shown the inside of a completed sheet metal door, including the wooden frame as well as the doorlatch *a*. The inside of the door is usually upholstered. Fig. 102 shows two other styles of doors which are made in a similar manner.

CONSTRUCTING AND DEVELOPING MOTOR HOODS

Another sheet metal part entering into the manufacture of the automobile is the motor hood, of which there are various styles and shapes. Six different styles or types of motor hoods will be described with the various rules for laying out the patterns by parallel lines, radial lines, and triangulation methods and will cover the principles required in laying out any other shape that may arise.

The first shape to be taken up is that of the regulation hinged hood, or sometimes called bonnet, which requires no pattern drafting. This type is simply a study in construction. Fig. 103 shows a regulation shaped motor hood in position. This hood is hinged at the top and sides. The hinges on either side and on the top allow the raising of either side, giving easy access to the motor. The constructive views are shown in Figs. 104 and 105. Fig. 104 shows a one-half vertical section through A B in Fig. 103, and Fig. 104 shows how the hinges are constructed. These hinged hoods are usually made of sheet steel and sometimes from

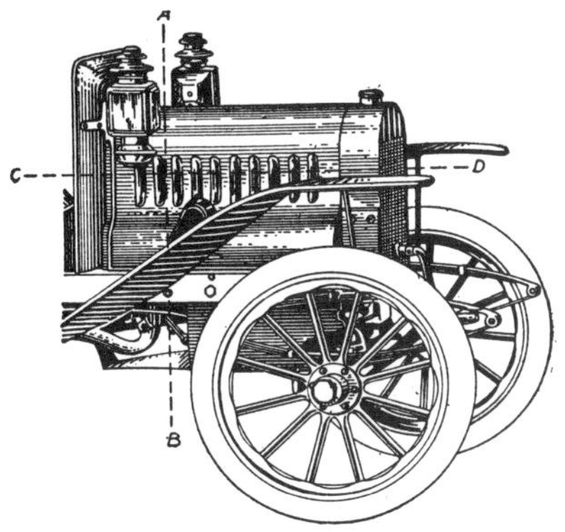

Fig. 103. Regulation Shaped Motor Hood in Position

Fig. 105. Horizontal Section through C D in Fig. 103

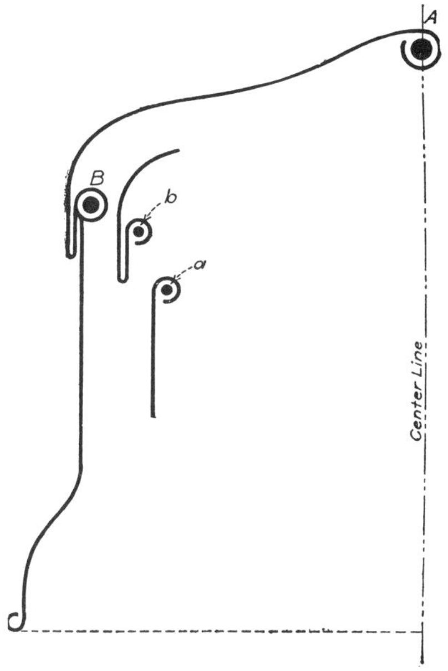

Fig. 104. Half Vertical Section through A B in Fig. 103

sheet aluminum. The pivot of the hinge is a ⅜ in. thick cold rolled rod, one at A and another at B. The hinge at B is so constructed that the lower part at *a* sets directly over *b* as shown at B. Fig. 105 shows a horizontal section through C D

Fig. 106. Angles on Dashboard and Radiator to Receive Hinged Hood

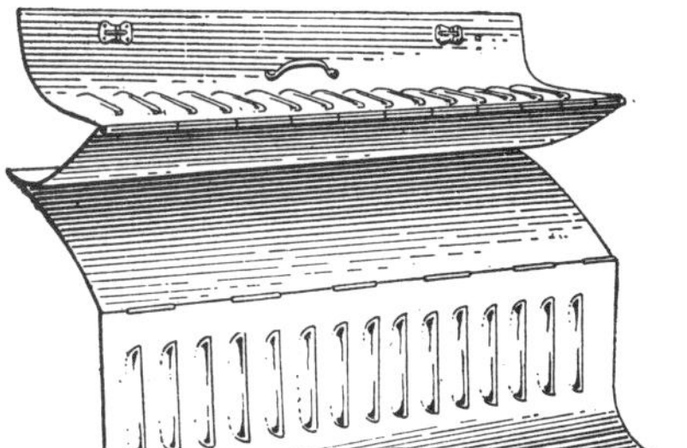

Fig. 107. The Hinged Hood Opened

in Fig. 103 and shows in Fig. 105 how a doubled hem edge is folded at *a* and *b*, which is bolted at one end to the dash *a* in Fig. 106 and at the other end to the radiator *b*. The openings along the side of the hood on C D in Fig. 103 are cut and stamped out by the power press, their section being shown in Fig. 105.

Fig. 107 shows how the regulation hinged bonnet or hood may be opened. Fig. 108 shows four different shapes of motor hoods. A B and C show those of the hinged type, while D shows a French type of hood, in which the beveled

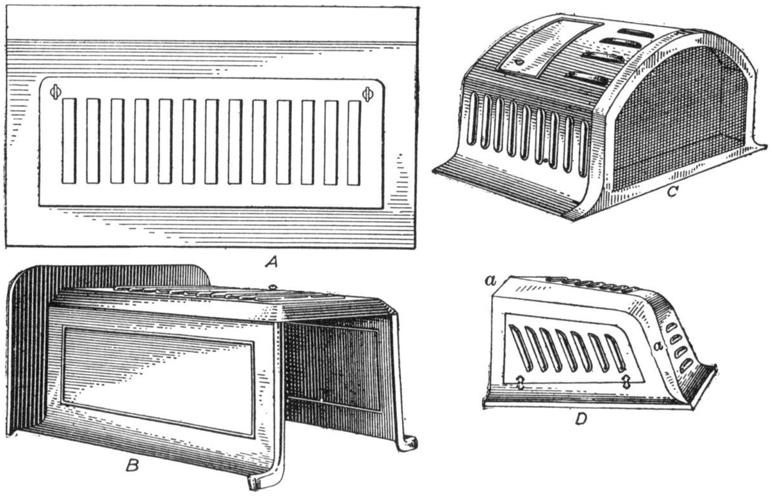

Fig. 108. Four Different Styles of Motor Hoods

corner *a a* is usually made of polished brass. The shape of the radiator usually gives the shape of the motor hood so that no standard shape of hood can be given. In the five types of hoods which will follow, pattern drafting becomes necessary in their development, and the next problem will take up one of the simple forms first.

Pattern for Tapering Square Molded Motor Hood

In Fig. 109 A shows the flat side and B the molded front of a tapering square molded motor hood. This form of hood may be developed by parallel lines as shown in detail in Fig. 110. In this figure 1 7 8 A shows the side elevation of the hood and 1′ 8′ B C the plan view. Note that the front from 1 to 7 in the side elevation is molded, while the plan view is tapering. The first step is to divide the molded part in the side elevation into equal spaces as shown from 1 to 7, from which points parallel to 8 A, project lines into the plan as shown intersecting the sides 1′ 8′ and C B. Now at right angles to 8′ B draw the line E D, on which place the girth of the profile 1 to 8 in the side elevation as shown by similar numbers on D E. Through these small figures at

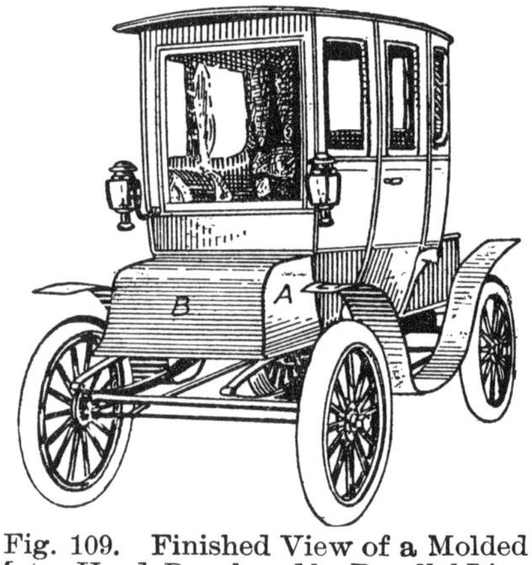

Fig. 109. Finished View of a Molded Motor Hood, Developed by Parallel Lines

right angles to D E draw the usual measuring lines, which intersect by lines drawn parallel to D E from similar numbered intersections on the sides 1′ 8′ and C B. A line traced through points thus obtained, as shown by F G H J will be the pattern for the front and top of the motor hood.

The pattern for the sides of the hood is obtained by taking the various divisions on the side line 1′ 8′ in the plan and placing them on the line 1 A extended in the side elevation, as shown by similar numbered divisions on K L. Through these small figures and at right angles to K L erect lines, which intersect by lines drawn parallel to K L from similar numbered intersections in the side elevation. Trace a line through points thus obtained, as shown by K L M, which will be the pattern for the sides of the hood. Flanges should be allowed for either seaming, riveting, or wiring.

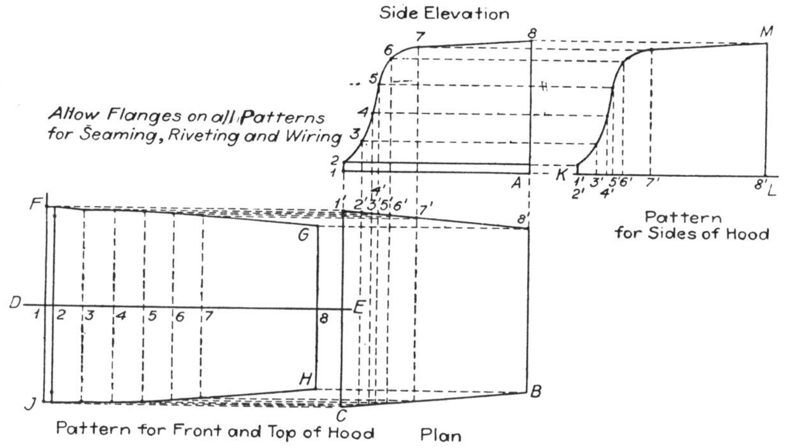

Fig. 110. Side Elevation, Plan, and Patterns for Motor Hood Shown in Fig. 109

Pattern for Flaring Piece in a Semi-Circular Motor Hood

Fig. 111 shows a finished front view of a semi-circular motor hood, in which a flaring piece is required with a seam at *a*. As this flaring piece represents a frustum of a right cone, it is developed by radial lines as shown in Fig. 112, in which A B C D E F represents the side elevation of the hood, and H 1 6 J K L the plan view. The horizontal section on the line A F in elevation is shown by J K L, while H 1 6 shows a similar section on the line B E in elevation, the semi-circles in plan being struck from the center *a*. Through *a* draw the vertical line 5 *b* and the horizontal line 1 *a*. Extend the bevel or flare B A in elevation until it intersects the vertical line *a b* at *c*,

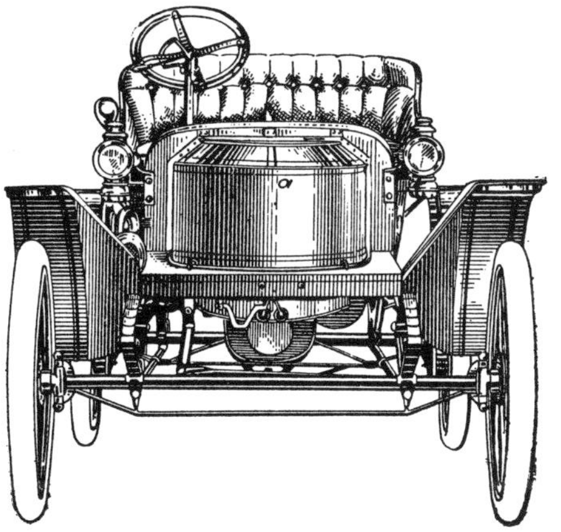

Fig. 111. Another Type of Semi-Circular Hood Requiring the Radial Line Method in Developing the Flaring Piece

which then represents the center with which to describe the pattern for the flaring strip as follows: Using c as center and c B, and c A as radii describe the arcs B 5 and A 5′. From any point as 1 on the arc B 5 draw the radial line to the center c cutting the inner arc at 1′. Now divide the half outer plan into equal parts, as shown from 1 to 5 to 6, and starting from point 1 in the pattern set off the girth of 1 to 5 in plan, as shown from 1 to 5 on the arc B 5 in the pattern. Draw a line from 5 to c, cutting the inner arc at 5′. Perpendicular to 5 c in the pattern, draw the lines 5 6 and 5′ 6′ equal to the distance 5 6 in the plan. Draw a line from 6 to 6′ in the pattern. Then will 1 6 6′ 1′ be the half pattern for the beveled or flaring strip, to which flanges must be allowed for riveting and seaming.

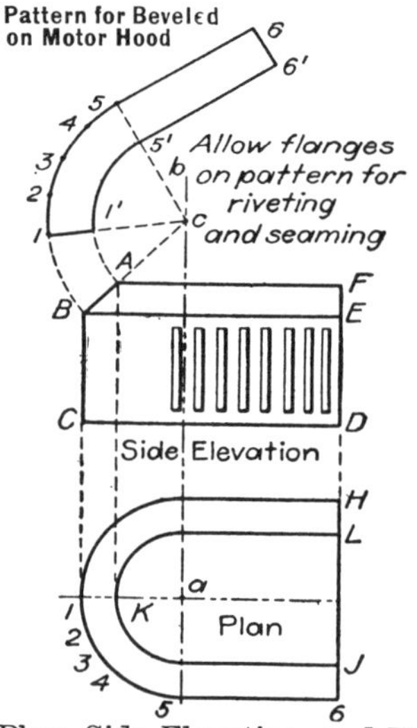

Fig. 112. Plan, Side Elevation, and Half Pattern for Beveled Strip on Semi-Circular Motor Hood Shown in Fig. 111

Patterns for Irregular Shaped Motor Hood

Another type of motor hood is shown in Fig. 113. It will be noticed that its top is part of a circle and is pitched; the sides are flaring while the front head is concave. This type of hood presents an interesting problem in pattern drafting and is shown in detail in Fig. 114. In this figure 1′ 1° e' d shows the side elevation of the hood, while B A 1 4 e^{v} shows the true profile or section on the line 1° e' in the side elevation. This pattern may be developed by parallel lines, but before this can be done it becomes necessary to find the true profile of the hood at right angles to 1′ 1° in the side elevation or on the line H G. Through the true profile on the line 1° e' draw the center line C D, and divide the top curve into equal parts, as shown from 1 to 4. From these points at right angles to D C, draw lines intersecting the line 1° e', as shown from 1° to 4°, from which points parallel to 1° 1′, draw lines indefinitely cutting the curve of the hood at 1′ 2′ 3′ and 4′. Divide the balance of the curve from 4′ to c into any desired number of spaces, as shown in this instance by a and b. From the intersections a, b, c, and d, draw lines parallel to 1′ 1° until they intersect the rear line of the hood at a', b', c', and d', and from these intersections lines are carried at right angles to D C until they cut this line and also intersect the profile at a^{v}, b^{v}, c^{v}, and d^{v}. Extend the lines drawn through *a*, *b*, *c*, *d* and e' indefinitely to the left as shown, and at right angles to these lines draw the line E F as shown. Measuring from the line D C in the true profile V take the various projections to points 2 to 4 to e^{v} and place them on similar numbered or lettered lines, measuring on either side of the line E F in X. Trace a line through points thus obtained as shown. Then will e 4 1 e'' be the true profile through H G in the side elevation.

This true profile X is used in obtaining the girth when developing the pattern for the side of the hood as follows: At right angles to 1' 1° draw the line H J or extend the line G H, as H J, on which place the girth of the half profile in X from 1 to 4 to *e*, as shown by similar numbers and letters on J H. Through these small figures and letters, and at right angles to J H draw lines, which intersect by lines drawn parallel to H J from similar numbered and lettered intersections on the rear line 1° e' and the front line 1 *c d* in the side elevation.

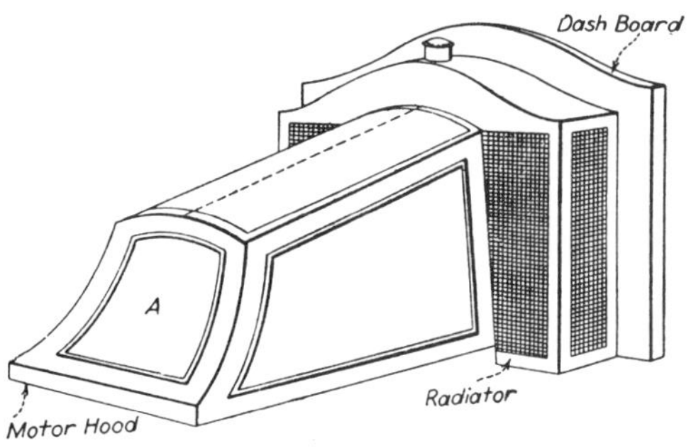

Fig. 113. View of an Irregular Shaped Motor Hood

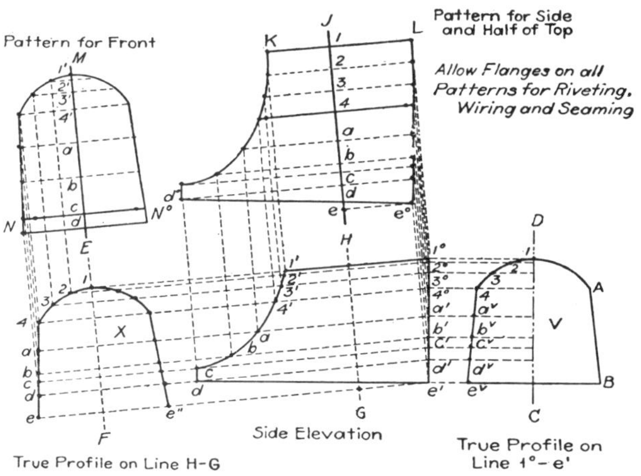

Fig. 114. Side Elevation, True Profiles, and Patterns in an Irregular Shaped Motor Hood Shown in Fig. 113

A line traced through points thus obtained, as shown by K L $e°$ $d°$ will be the pattern for the side and half of the top of the motor hood.

To obtain the pattern for the front piece shown by A in Fig. 113, extend the line F E in Fig. 114 as shown by E M, on which place the girth of the curve 1' *c d* in the side elevation as shown by similar letters and figures on E M. Through these small letters and figures, and at right angles to E M draw lines as shown, which intersect by lines drawn parallel to F E from similar intersections in the profile X. A line traced through these points of intersection from M to N is the half pattern. Trace this half opposite the line M E. Then will N M N° be the full pattern for the front head of the hood. Flanges must be allowed on all patterns for riveting, wiring, and seaming.

Pattern for French Type of Motor Hood

In Fig. 115 is shown a perspective view of a French type of motor hood. Note its peculiar shape and the chamfer along the corners of the hood. This style of hood usually has a brass handle riveted at *a* and is hinged along the top at *b*. By raising the handle at *a* the hood may be lifted with ease. Fig. 116 shows a different type of car with a front view of a French type of motor hood with the curves in the front and top A, in the sides B, also the chamfers J and J, which are usually made of polished brass. The method of developing the patterns for this style of hood is shown in Fig. 117, in which parallel lines and triangulation are employed. Not alone must care be taken in the development of the patterns, but the plan, section, and elevation must be accurately drawn as follows:

Fig. 115. Perspective View of a French Type of Motor Hood

Fig. 116. Front View of French Type of Motor Hood, Showing Chamfer, also Curves in the Top, Front, and Sides

First draw the outline of the plan as shown by D C 6 6′ giving the desired taper as shown. Through the plan draw the center line K J. Now in its proper position and in line with the plan draw the outline of the side elevation of the hood, as shown by A B 11 6, being careful to draw graceful curves from 10 to 6. From C in the plan draw the vertical line C *b*. Using C as a center and C B as radius, draw an arc intersecting the vertical line C *b* at E. From E draw the horizontal line E H equal in length to C D in plan. Again using C as a center and C 11 in the side elevation as radius, draw an arc intersecting the vertical line C *b* at 11, from which point draw the horizontal line 11 G. Between the horizontal lines 11 G and H E draw the profile or section of the dash hood, as shown by E H G 11°, which will represent the section on the line D C in plan. From 1° in the lower end of the chamfer draw a horizontal line cutting the vertical line *b* E at 1. Using C as a center and C 1 as radius, draw the quarter circle intersecting the rear line B 11 in the side elevation at 1. From this point 1 draw the lower line of the chamfer, making a graceful taper to the point 6 as shown. When the outline of the chamfer 1 to 6 and 6 to 11 in the side elevation is being drawn the taper should be gradual,

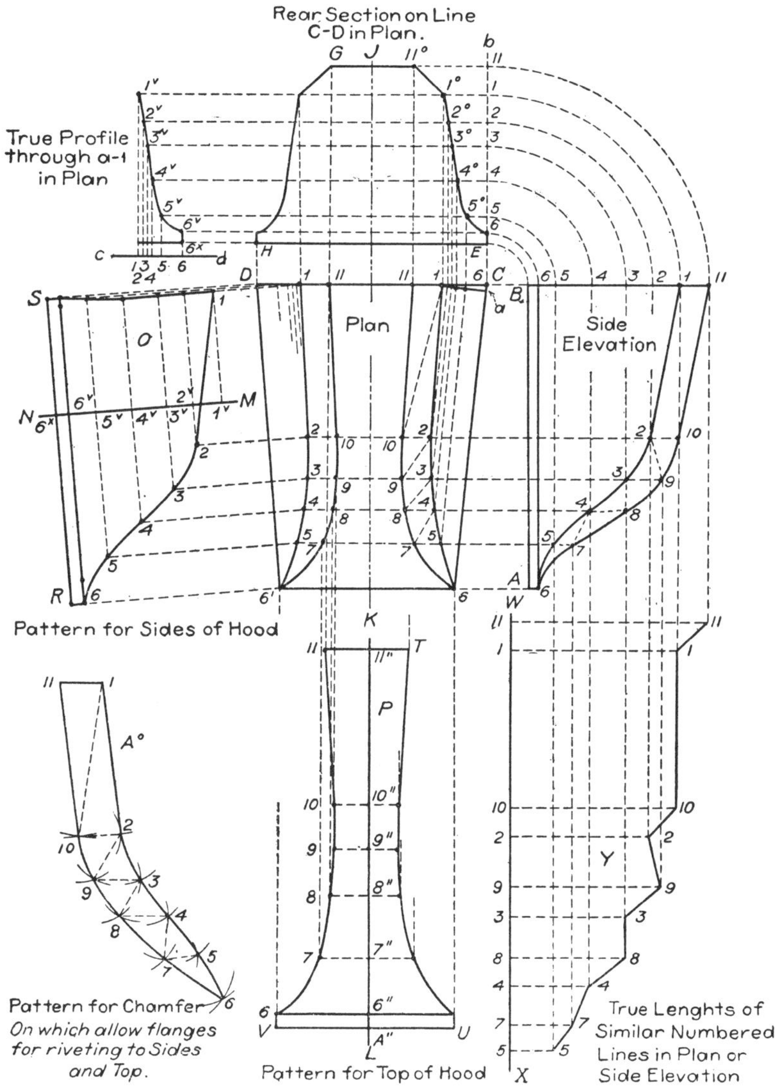

Fig. 117. Plan, Side Elevation, True Sections, and Patterns for French Type of Motor Hood

as these lines when once drawn remain fixed and cannot be changed when developing the pattern.

While the miter line between the lower edge of the chamfer and side of the hood, shown by 1 6 in the side elevation, has been drawn at pleasure, this same miter line shown by 1 6 in the plan cannot be drawn at will but must be projected as follows: Divide the curved part from 2 to 6 in the side elevation into equal parts as shown by points 2, 3, 4, 5 and 6, from which draw lines parallel to A B until they cut the line B 11 from 1 to 6 as shown. Now using C as a center and with radii equal to C 5, C 4, C 3, and C 2 draw quadrants cutting the vertical line C *b* from 5 to 2 as shown. From these points 2 to 5 on C *b* draw horizontal lines cutting the profile 11° E from 2° to 5°. From these intersections 6 to 11° in the rear section drop vertical lines cutting the line C D in plan as shown. From the intersections between 1 and 6 on the line D C draw lines parallel to C 6 and intersect same by lines drawn from similar numbers in the side elevation at right angles to A B, thus obtaining the intersections 1 to 6 in the plan view. A line traced through these points as shown will be the miter line in plan between the lower end of the chamfer and the side of the hood, shown by 1 6 in the side elevation. To obtain the plan line of the line shown from 6 to 11 in the side elevation draw a graceful, tapering curve from the point 11 in plan to the corner 6. Trace these miter lines 1 to 6 and 6 to 11 in plan opposite the center line as shown. While the full rear section and plan are here shown, in practical work only one-half is necessary.

The next step is to find the true section of the side of the hood at right angles to C 6 in plan. At right angles to C 6 and from the point 1 draw the line 1 *a* crossing the lines previously drawn as shown. Draw any horizontal line as *c d*, on which place the various divisions contained in 1 *a* in plan, as shown by similar numbers on *c d*. From these numbers erect vertical lines, which intersect by horizontal lines drawn from similar numbers in the rear section. A line traced through these points as shown from 1^{v} to 6^{v} to 6^{x} will be the true profile through *a* 1 in plan.

The pattern for the sides of the hood is now in order, for which proceed as follows: At right angles to D 6′ in plan draw any line as M N, on which place the girth of the true profile through *a* 1 in plan as shown by similar numbers on N M. At right angles to M N through the small figures 1^{v} to 6^{x} draw lines as shown, which intersect by lines drawn at right angles to D 6′ in plan from similar numbered intersections on the miter line 6′ 1 and the rear line 1 D in plan. Trace a line through points thus obtained. Then will 1 6 R S be the pattern for the sides of the motor hood, to which flanges must be allowed along R S and S 1 for stiffening purposes.

The pattern for the top and front of the hood is next in order. Instead of dividing the curved part from 10 to 6 in the side elevation into equal spaces, simply extend the lines already drawn through points 2, 3, 4, and 5 and obtain points 10, 9, 8, and 7 respectively in elevation and plan. Now take the girth of the profile 11 6 A in the side elevation and place it on the center line in plan extended as K L, as shown by similar numbers 11″ to A″. Through these points and at right angles to K L, draw lines, which intersect by lines drawn parallel to K L from similar numbered intersections on the miter line 6 to 11 and 6′ to 11 in plan. A line traced through points thus obtained as shown by T U V 11 will be the desired pattern. A flange is allowed along U V for stiffening purposes and a heavy brass hinge riveted along the top at T 11.

The last pattern is for the chamfer, which is developed by triangulation as follows: Connect opposite corners in plan and elevation as from 1 to 10, 2 to 9, 3 to 8, and 4 to 7. These lines in plan then represent the bases of triangles, the altitudes of which will be equal to the various horizontal distances between similar numbered points in the side elevation and are obtained as follows: Draw any line parallel to B A in the side elevation as W X on which place the various distances in plan as 11 to 1 to 10 to 2 to 9 to 3 to 8 to 4 to 7 to 5 as shown by similar numbers on W X. From these points and at right angles to W X erect lines, which intersect by lines drawn parallel to W X from similar numbers in the side elevation, thus obtaining the true lengths as shown by the heavy lines 11 to 5 in Y. These lines then represent the true lengths of the lines shown by similar numbers in plan. The true lengths of the curved lines 11 to 6 and 6 to 1 in plan are shown respectively from 11 to 6 in the pattern P and 6 to 1 in the pattern O. Take the distance from 11 to 1 in the diagram Y or 11° to 1° in the rear section, both of which are similar and place it as shown from 11 to 1 in the chamfer pattern A°. Now with 11 10 in the pattern P as radius and 11 in A° as center describe the arc 10, which intersect by an arc struck from 1 as center and 1 10 in diagram Y as radius. With 1 2 in the pattern O as radius and 1 in A° as center describe the arc 2, which intersect by an arc struck from 10 as center and 10 2 in Y as radius. Proceed in this manner, using alternately, first the divisions in the miter cut in P, then the true lengths in Y; the proper division in O, then again the true length in Y until the line 7 5 in the chamfer pattern A° has been obtained. Next with 7 6 in P as radius and 7 in A° as center draw the arc 6, which intersect by an arc struck from 5 as center and 5 6 in the pattern O as radius. Trace a line through these points and then will 1 6 11 be the pattern for the chamfer, one of which should be bent right and one left. Flanges must be allowed on this chamfer pattern for riveting to the sides, top, and front.

Pattern for French Motor Hood without Chamfer

In this problem will be taken up another type of French motor hood without a chamfer as shown in Fig. 118. These hoods are usually made from one piece, turned over along *b c*, the curved miter being reinforced by riveting to a brass angle *a* to *b*. Along the top a brass hinge is riveted as *c d* and at the bottom a brass handle, *e*, is riveted for lifting the hood. A flange *f* is allowed along the bottom for stiffening purposes. The method of developing a hood of this kind is shown in detail in Fig. 119 in which D C 7′ 7″ shows the base outline of the hood and A B 1 7 the side elevation placed in its proper position. Parallel and equal in length to D C in plan draw the line 7° 7^{v} making the distance C 7° equal to C B. From C erect the vertical line C *b*, and using C as center and C 1 in the side elevation as radius, draw the arc cutting C *b* at 1. From this point 1 draw a horizontal line to the left indefinitely and draw the desired profile or section on the line D C in plan.

The plan, side elevation, and rear section having been drawn the next step is to find the miter line in plan between the sides and top and front of the hood. This is done as follows: Divide the curved part in the side elevation from 2 to 7 into equal parts as shown, from which points and parallel to A B draw lines intersecting the rear line B 1 as shown by similar numbers. Now using C as center with the various numbers on B 1 as radii draw arcs intersecting the vertical line

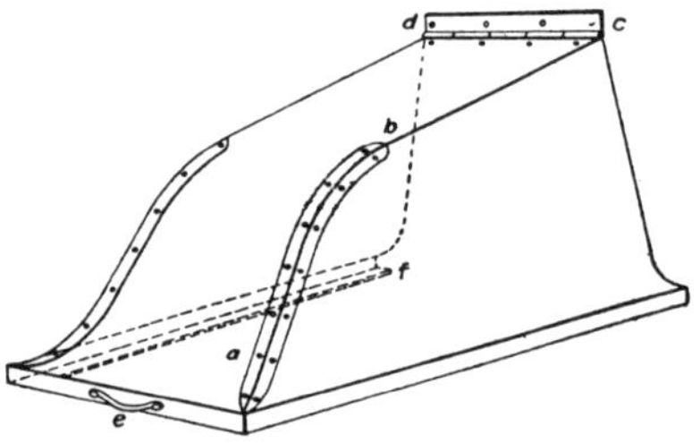

Fig. 118. Another Type of French Hood without the Chamfer

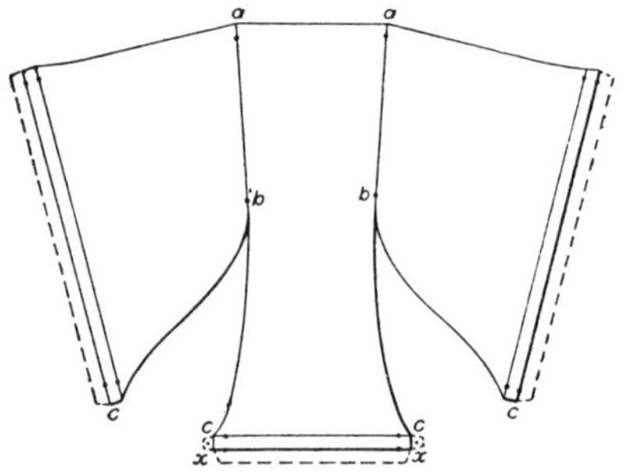

Fig. 120. Hood Developed in One Piece

C *b* also from 1 to 7, from which points and parallel to 7° 7^{v} draw lines to the left indefinitely, cutting the rear section from 1° to 7°. Between 6° and 7 establish another point *e*° because the curve is too deep, and project the point back to *e* on the line C *b* and thence to the curve in the side elevation. From the various intersections 1° to 7° in the rear section drop vertical lines to the line D C in the plan from which points draw lines parallel to C 7′ in plan and intersect them by lines drawn at right angles to A B from similar numbered points in the side elevation, thus obtaining the miter line shown from 1 to 7′ in the plan. Trace this line on the opposite side in the plan as shown from 1 to 7″. In practice only the one-half plan and rear section are necessary.

While 1° to 7° in the rear section shows the true profile on D C in plan, another profile must be found at right angles to the side of the hood as follows: From 1 in plan and at right angles to C 7′ draw the line 1 *a* cutting the lines previously drawn. Take the various projections on this line 1 *a* and place them on the horizontal line *c d*, as shown by 1, 2, 3, 4, 5, 6, *e*, and 7. From these points draw perpendicular lines to *c d*, intersecting horizontal lines of similar numbers previously drawn through the rear section. A line traced through these points 1 to 7^x will give the true profile through 1 *a* in plan.

For the pattern for the sides take the girth of the true profile from 1 to 7^x and place it on the line H J, drawn at right angles to D 7″ in plan as shown by similar numbers. Through these small figures and at right angles to J H draw lines,

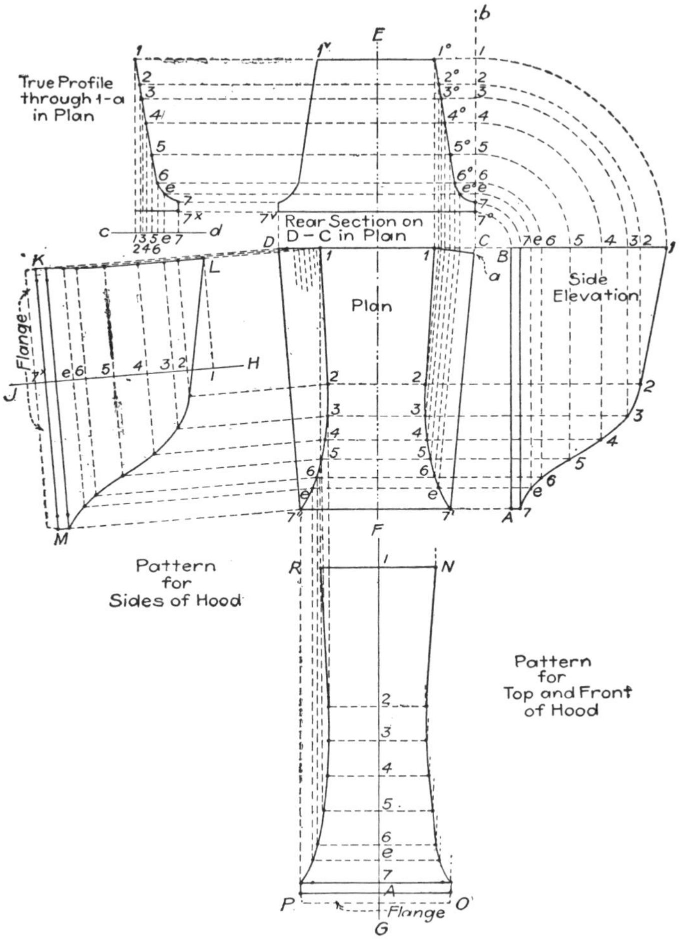

Fig. 119. Plan, Side Elevation, True Section, and Patterns for French Type of Motor Hood without Chamfer

which intersect by lines drawn parallel to J H or at right angles to D 7″, from similar numbered intersections in the miter line 1 7″ in plan and 1 D on the rear line D C. A line traced through these points of intersection as shown by K L M will be the pattern for the sides of the hood, to which a stiffening edge has been added as indicated by the dotted line.

The pattern for the front and top is obtained by taking the girth from 1 to 7 to A in the side elevation and placing it as shown by similar numbers on the center line E F in plan extended as F G. Through these small figures and at right angles to G F draw the usual measuring lines, which intersect by lines drawn parallel to F G from similar points of intersection in the miter lines in plan as shown. After a line has been traced through points thus obtained as shown by N O P R this will be the pattern for the top and front of the hood, to which the stiffening flange P O has been added.

As previously mentioned this style of hood is made up in one piece by joining the front pattern to the two sides as shown in Fig. 120 in which the full pattern is shown. When making the bend along *a b* this is rounded slightly at *a* then runs tapering and to a sharp corner at *b*. From *b* to *c* the corners have raw edges, but with a flange at *x x* for riveting. From *b* to *c* the corners are reinforced with the brass angles *a b* shown in Fig. 118.

PATTERN AND CONSTRUCTION OF FRONT WHEEL FENDERS AND SPLASH GUARDS

In Fig. 121 is shown a view of a typical front wheel fender and splash guard usually called the front mud guard. This mud guard consists of a fender, rim and filler or splash guard. The splash guard is usually fastened along the top of the angle of the chassis as shown at A and is bolted to the running board filler or mud guard at B. When the fender is pressed by hand the rim is seamed or riveted to the fender, as will be explained later. If the fender is pressed by machine the rim is usually pressed on to the fender. Whether the rim is joined to the fender by hand or by the press, the filler or splash guard must always be double seamed separately, as will be explained in the constructive part. There are various types of fenders and in this part will be described seven different types, requiring radial lines as well as triangulation in the development of the patterns. In drawing the detail of the fenders, large graceful guards should be used which come over the wheels to check the flying dirt entirely. There should be no open space between the frame and fender for dirt to fly through or dust to rise. Mud shields or guards should be provided between the running board and the angle frame of the chassis which give protection from flying mud and dirt. How these running board mud guards are constructed and fastened will be explained in the constructive part, after the first pattern has been developed.

Pattern for Fender Joining Channel Frame and Running Board Mud Guard

The first pattern to be developed will be that of the front wheel fender of the type shown in Fig. 121, requiring triangulation in the development of the pattern as shown in Fig. 122, in which *b* 1 6 7 8 11 shows the side view of the fender and mud guard and *h i* 8 11 6′ 2 the end elevation. In the side elevation B shows the channel of the chassis, a section of which is shown by A. In the side elevation 1 6 7 8 shows the profile of the fender also shown in the end view by 1 *h i* 8. In the side elevation 6 1 *b* shows the side elevation of the filler or splash guard which is shown in the end elevation by 2 6′ 6 while 7 8 11 in the side elevation

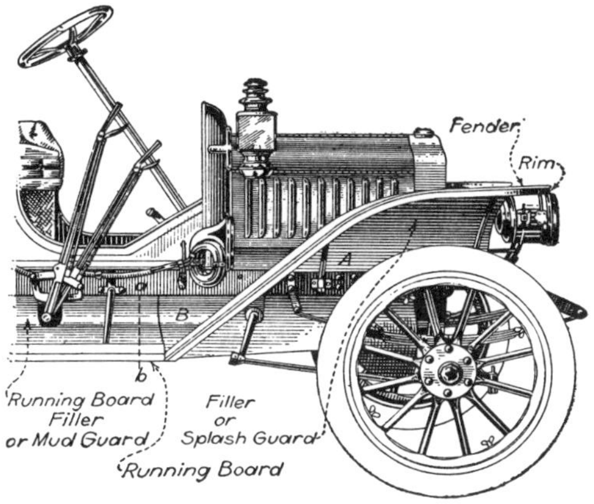

Fig. 121. View of a Typical Fender and Splash Guard Joining Frame and Running Board Mud Guard

Fig. 123. Front View of Fender and Splash Guard Showing Connection between Splash Guard and Rim

shows that part of the splash guard connecting to the bottom of the channel along 7 11, to the fender along 7 8, and to the running board filler or guard along 8 11. This guard below the channel frame is shown in the end view by 7 7′ 11 8. The top of the fender is stiffened and finished by placing two swedges or beads in it as shown in the end view and in the part plan of the top above the side elevation. It will be noticed that the circular head of the fender *o r n* is not shown in the side elevation as this is not necessary in the development of the filler.

The side and end elevations having been drawn in their proper relative positions, the first step is to divide the curved part of the profile of the fender into equal parts as shown by the small figures 1 to 6 to 7 to 8. Draw a line from 1 to *b* in the side elevation, and from points 1, 2, 3, 4, and 5 draw lines to 6. Divide the curve of the running board filler from 8 to 11 in the end elevation into equal spaces as shown by 9 and 10, from which points draw horizontal lines to the side elevation as shown by 9 and 10, and from these points draw lines to the corner 7. From similar points 8, 9 and 10 in the end elevation draw lines to 7′ or to the bottom of the channel A, representing the point 7 in the side elevation. From points 1 to 6 in the side elevation project lines to the end elevation as shown by similar numbers, and from these points 1 to 6 draw lines to 6′ or the top of the channel which represents 6 in the side elevation. The various dotted lines in the side elevation then represent the bases of triangles which will be constructed, the heights of which will be equal to the horizontal distances between similar numbered points in the end elevation. For example: To obtain the true lengths of the lines 6 2 and 7 10 in the side elevation, take these distances and set them off on the vertical line dropped from the frame line 6′ 7′ in the end elevation as shown by 6 2′ and 7′ 10′ in the diagram C. At right angles to these lines and from points 2′ and 10′ draw lines, which intersect by vertical lines dropped from intersections 2 and 10 in the end elevation, thus obtaining points 2 and 10 in diagram C. Draw lines from 2 to 6 and 10 to 7′ in C, which give the true lengths of similar numbered lines in the side elevation. In this manner obtain all the true lengths of the lines shown in the side elevation as shown by similar numbers in the diagram C.

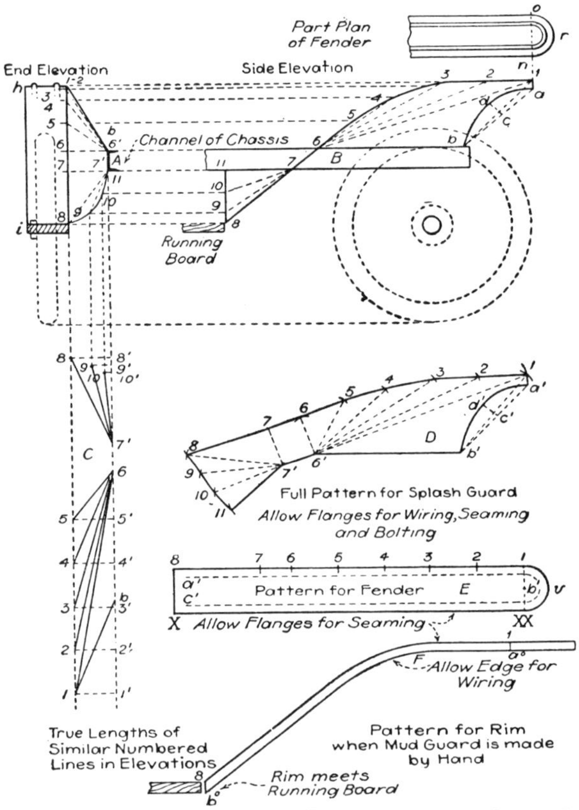

Fig. 122. Elevations, Part Plan, True Lengths, and Patterns for Mud Guard, Shown in Fig. 121

The pattern for the filler or splash guard is now in order as shown in diagram D. Take the distance from 6 to *b* in the side elevation and place it as shown from 6′ to *b*′ in D. With radius equal to *b* 1 in diagram C and *b*′ in D as center describe the arc 1, which intersect by an arc struck from 6′ as center and 6 1 in the true lengths C as radius. Now with radii equal to 6 2, 6 3, 6 4, and 6 5 in C and 6′ in D as center, draw the arcs 2, 3, 4, and 5. With 6′ 6 in the end elevation as radius, and 6′ in D as center draw the arc 6. Set the dividers equal to the spaces between 1 and 6 in the side elevation and starting from 1 in the pattern D step to arc 2, then to 3, and so on until the point 6 is obtained. Draw a line from 6 to 6′, at right angles to which draw the lines 6 7 and 6′ 7′ equal to 6 7 in the side elevation. Then draw a line from 7 to 7′ in the pattern D. With 7 8 in the side elevation as radius and 7 in the pattern D as center, draw the arc 8, which intersect by an arc struck from 7′ as center and 7′ 8 in the true lengths C as radius. With radii equal to 7′ 9 and 7′ 10 in C and 7′ in D as center draw the arcs 9 and 10. Using 7 11 in the side elevation as radius, and 7′ in D as center, draw the arc 11. Set the dividers equal to the divisions between 8 and 11 in the end elevation, and starting from the point 8 in the pattern D step to arc 8, then to 9, 10, and 11 as shown.

To reproduce the curve shown by *a d b* in the side elevation on to the pattern draw a line from *a* to *b*, which bisect and obtain *c*. At right angles to *a b* from the point *c* draw the line *c d* meeting the curve at *d*. Take the distance from 1 to *a* or the width of the rim, and place it as shown by 1 *a*′ in D at right angles to 2 1.

Draw a line from *a*′ to *b*′, which bisect and obtain *c*′. From *c*′ and at right angles to *a*′ *b*′ draw the line *c*′ *d*′ equal in length to *c d* in the side elevation. Draw a curved line through *a*′ *d*′ *b*′ in D and trace a line through the various points 1 to 8 to 11, which will be the true pattern for the filler or splash guard. Edges should be allowed on the pattern from 1 to 8 for seaming to the fender; from 8 to 11 for bolting to the running board filler, the profile of which is shown from 8 to 11 in the end elevation; from 11 to *b*′ in D for bolting to the channel of the chassis and from *b*′ to *d*′ to *a*′ for wiring.

The end elevation of the fender is shown by *h i* 8 2 and its pattern is obtained as follows: Draw any line as 8 1 in E, on which place the girth of the profile of the fender 8 1 in the side elevation as shown by similar numbers in E. Draw the perpendicular 8 X in E equal to *i* 8 in the end elevation, and complete the rectangle 1 8 X XX. Bisect 1 XX and obtain *b*′, which is the center for completing the semicircular head of the fender shown in the part plan by *n r o*. The positions where the beads will be placed on the fender are shown by *a*′ and *c*′ in E. On this pattern E flanges must be allowed for seaming to the splash guard and rim.

When the top fender is not pressed by machine the rim is laid out separately by taking a reproduction of the profile 8 1 in the side elevation and placing it as shown by 8 1 in F. Take the distance 1 *a* in the side elevation and place it as shown by 1 a° in F, and from a° draw a line parallel to the curve 1 8 as shown by a° b°. Add to 1 a° the distance o°, which is the girth of *n r o* in the part plan or the girth of XX *v* 1 in the pattern E. Along 1 8 in the pattern F allow an edge for seaming to the fender top, also an edge along a° b° for wiring.

When the mud guard is being seamed together the pattern for the rim meets the pattern of the splash guard, as shown in Fig. 123, where the splash guard A meets the rim V as shown by the arrow on the right side of the illustration.

Constructing the Mud Guard

Fig. 124 shows a completed guard ready to be attached to the automobile. Note that the curve *a b* is that part which joins the running board filler at B in Fig. 121. A slight bend is made along *b c* in Fig. 124, and is shown in the pattern

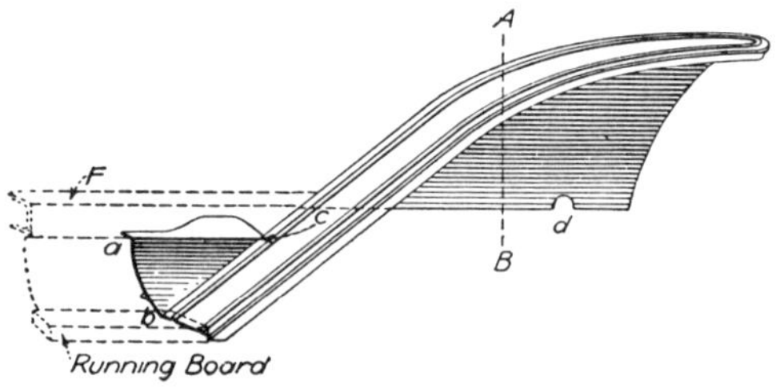

Fig. 124. View of Finished Front Wheel Mud Guard

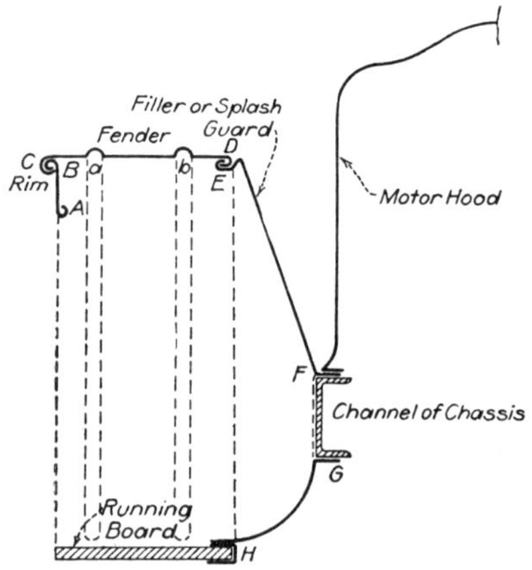

Fig. 125. Construction of Mud Guard when Made from Separate Pieces, also Showing Section at *a b* and A B respectively in Figs. 121 and 124

in Fig. 122 by 7′ 8. The hole *d* in Fig. 124 is punched in the flat by machine, to allow the iron brackets which support the fender to pass through. One of these brackets and sockets is shown by A in Fig. 121. In constructing the mud guard,

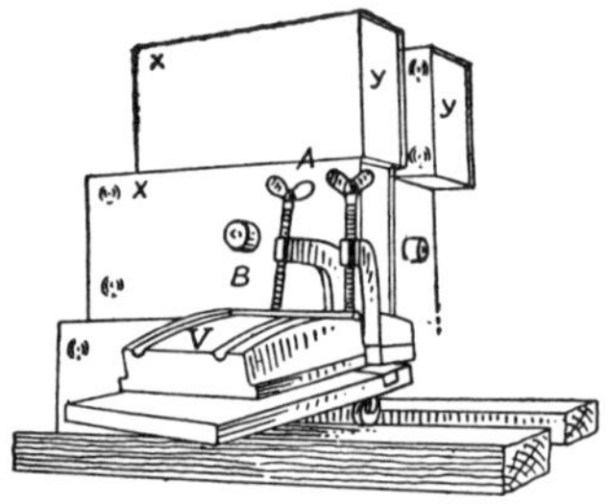

Fig. 126. Die Used in Bending Curves in Fender

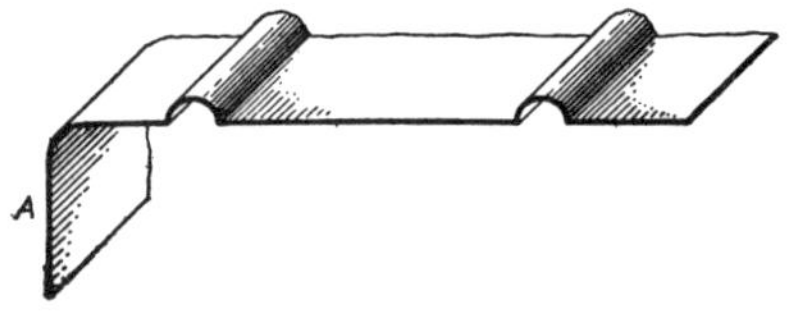

Fig. 127. Pressing the Fender and Rim in One Piece

sometimes the rim and fender are made from separate pieces or stamped or pressed in one. Fig. 125 shows the construction when the fender and rim are separate and also shows a section through *a b* and A B respectively in Figs. 121 and 124. In the Fig. 125 A shows the rim with a wired edge at A and B. This is done on the small turner machine. The fender is formed up as follows: After the pattern for it has been laid out as previously described, the beads or swedges are put in the metal on the flat with a No. 1 beader, using ½ or ⅝ in. beads. This beaded flat fender may now be formed on any large curved surface or passed through a set of male and female rolls, the profiles of which are similar to the section shown by B *a* *b* D, the same as a stove pipe former, there being a top and bottom roll and a rear roll, which may be raised or lowered to give the desired curve, and the rolls may be reversed whichever way the curve is desired.

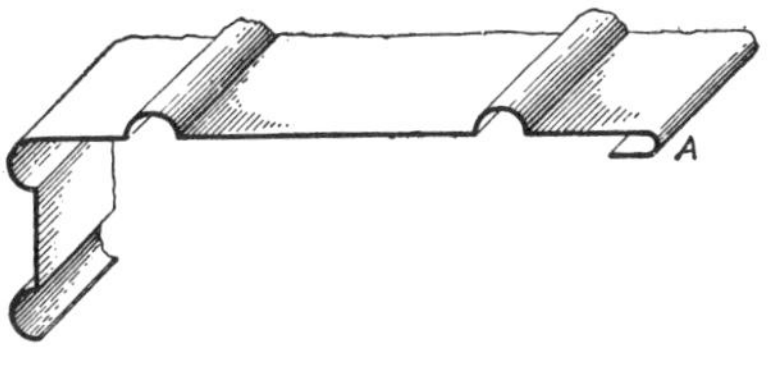

Fig. 129. Finished Section of Pressed Fender with Lock Attached

Another mold used to form the lower part of the fender where it meets the running board is shown in Fig. 126, in which V shows a cast-iron mold, with a clamp at B operated by the set screws A. In operating this die the fender is secured under the clamp B by fastening the set screws A and while in this position the metal is annealed or softened by heating that part to be curved with a brazing blow torch, and then pressing slowly but firmly until the proper curve has been obtained. When the fender has the proper curve the bead at C in Fig. 125 is made on the turner

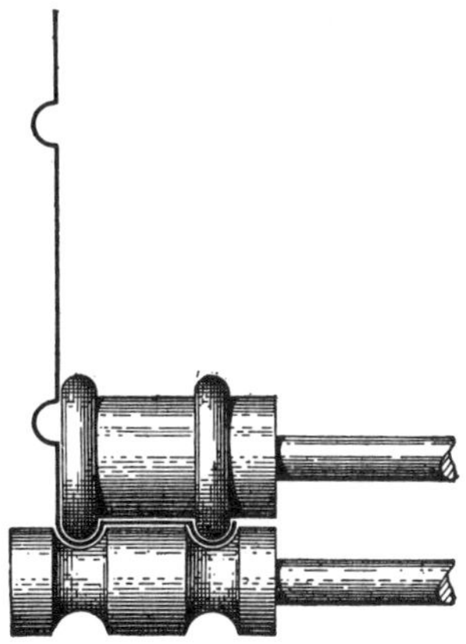
Fig. 128. Beading the Rim

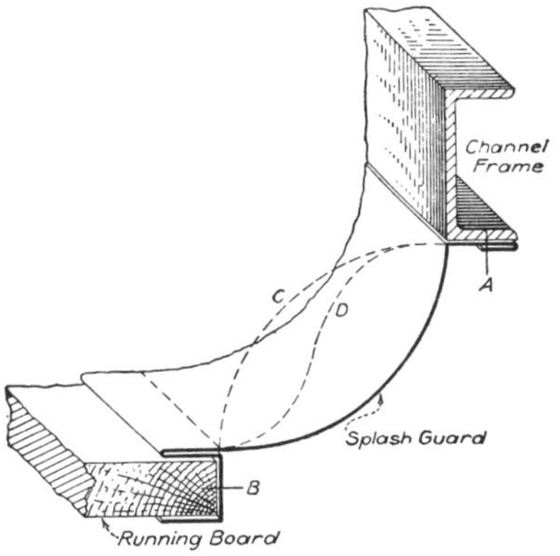

Fig. 130. Section Through *a* *b* in Fig. 121 Showing Construction of Running Board, Splash or Mud Shield

machine and hooks tightly around the bead B of the rim. At D a lock is made on the thick edge machine which locks to the splash guard at E. The lock on the splash guard at E is also made on the thick edge machine, after which the seam D is grooved along the inside at E, with a hand groover to get started, then the power groover is used to finish with. Flanges are allowed to the lower part of the splash guard to bolt to the channel of the chassis, at F and at G. That part of the splash guard from G to H joins the running board filler which will be described in regular order.

When the fender and rim is made in one piece it is pressed in a large power press which brings the rim in the position shown by A in Fig 127, after which the rim is passed through a set of male and female power swedges as shown in Fig. 128, which gives the rim a beaded finish as shown in Fig. 129. When this has been done the seaming lock at A is made on the thick edge machine and seamed to the splash guard as before described.

Fig. 130 shows the construction of the running board filler or splash guard. In this case the splash guard is concave as shown by the solid line. Sometimes it has the formation shown by the dotted lines, one being convex C and the other an ogee curve D whichever shape is used the constructive features are similar. A hem edge is placed at the top of the guard and the flange bolted to the bottom of the channel at A. At the bottom of the guard a groove is formed to same and slips over the running board as shown at B and is screwed as shown.

Pattern for Projecting Rim or Lip of Mud Guard

On most of the fenders the rim is usually of the same width throughout as indicated by V in Fig. 123. On some fenders the top is provided with a projecting rim or lip as indicated by x in Fig. 131. This lip is made separate and is

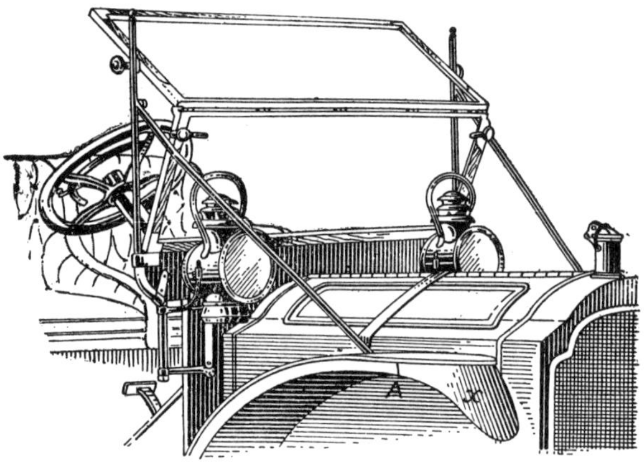

Fig. 131. View of Projecting Rim on Front Wheel Mud Guard

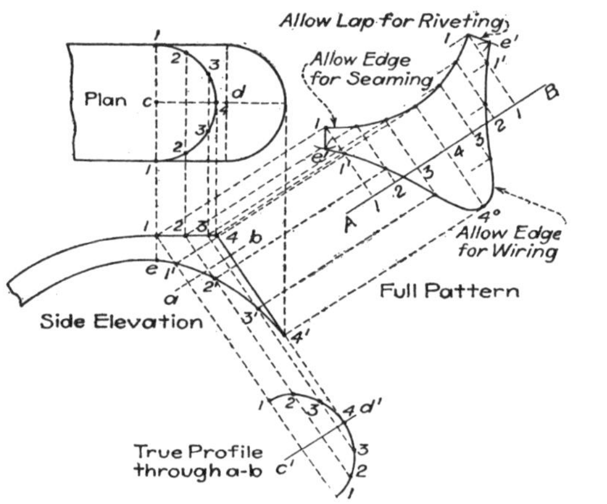

Fig. 132. Developing the Pattern for the Projecting Lip or Rim Shown in Fig. 131

joined and riveted to the fender at A on the outside, as well as where it joins to the inside splash guard or filler. While this lip x may be developed by triangulation, a simpler method may be employed by using parallel lines as shown in Fig. 132, in which the side elevation of the lip is shown by 1 4 4′ e and its plan directly above it as shown. The top view of the fender on the line 1 4 in elevation is shown by 1 4 1 in plan while the semi-circle struck from the point d in plan represents the view horizontally through the point 4′ in side elevation. The first step is to divide the semi-circle in plan struck from the center *c* into equal parts as shown by the small figures 1 to 4 to 1, from which points drop vertical lines until they intersect the top line of the fender 1 4 in the side elevation as shown by similar numbers. From these points and parallel to 4 4′, or the pitch of the lip, draw lines indefinitely as shown, cutting the lower line of the lip from 1′ to 3′. At right angles to the lines just drawn draw the line *c′ d′*. Now measuring from the line *c d* in plan, take the various projections to points 1, 2 and 3 and place these projections on either side of the line *c′ d′* on similar numbered lines as shown, resulting when a line is traced through them in the true profile through *a b* in the side elevation, or the true profile at right angles to 4 4′. This true profile is used to obtain the girth of the material required when developing the pattern.

For the pattern proceed as follows: At right angles to 4 4′ draw the line A B, on which place the girth of the true profile, as shown by similar numbers 1 to 4 to 1 on A B. Through these small figures at right angles to A B draw lines as shown, which intersect by lines drawn parallel to A B from similar numbered intersections 1 to 4 and 1′ to 4′ in the side elevation. Trace a line through points thus obtained as shown from 1 to 1 to 1′ to 4° to 1′ to 1 which will be the pattern for the lip. On either side of the pattern on the lines 1 1′ add the triangular piece 1 *e* 1′ in the side elevation. This is done by using 1 *e* in the side elevation as radius and with 1 in the pattern as centers, describe the arcs *e′* and *e′*, which intersect by arcs struck from 1′ as centers and 1′ *e* in the side elevation as radius. Connect lines in the pattern from 1 to *e′* to 1′. This line 1 *e′* meets the rim 1 *e* in the side elevation. Edges must be allowed for seaming, riveting and wiring as shown on the pattern.

Pattern for Mud Guard Joining Running Board Only

In Fig. 133 is shown another type of mud guard, which does not connect to the running board splash guard as in Fig. 121, but joins to the running board only as shown in Fig. 133. The splash guard below *c* joins to the channel of the chassis, from which it runs to the inner edge of the running board from *b* to *a*. It will be noticed that the rim projects as at *d*, the pattern for which was previously described. The only part which will be developed in this mud guard, is the splash guard or filler. The method of developing the fender and rim will be omitted here as this has already been described.

Fig. 134 shows how this splash guard or filler is developed. First draw the side elevation as shown, 1 6 representing the profile of the fender, 6 7 the end that meets the running board, 8 9 the intersection with the top of the channel, while a straight line connects 7 to 8. In its proper position to the left of the side elevation draw the end elevation as shown, the distance 5 to 9 showing the projection of the splash guard between the inner side of the fender and channel. Now divide the curved part of the fender into equal spaces as shown between 2 and 5 and draw lines from 2, 3 and 4 to 8. Also draw a line from 2 to 9 and 9 to 1. Draw the width of the rim 1 *a*, and draw the curve *a c* 9 at pleasure. Number the points in the side elevation from 1 to 9 as shown. From points 1 to 7 project horizontal lines to the end elevation, cutting the outer edge of the splash guard also from 1 to 7. In a similar manner project 8 9 on the channel in the side elevation to a similar channel in the end view, shown by 8 9. Connect similar numbers in the end view to correspond with similar number in the side. Then will the lines in the side elevation represent the bases of triangles which will be constructed in diagram A, the altitudes or heights of which are equal to the horizontal distances between similar numbers in the end elevation.

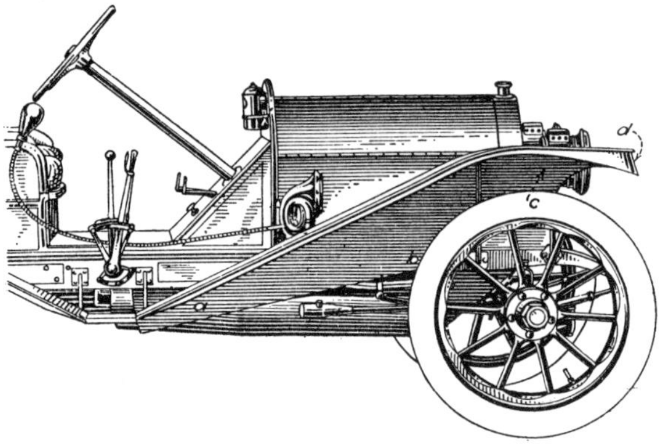

Fig. 133. Another Type of Mud Guard Joining Running Board Only

From points 7 and 8 in the end elevation drop vertical lines indefinitely as shown. Take the various lengths in the side elevation as from 1 to 9 to 2 to 8 to 3, 4, 5, 6 and 7 and place them as shown by similar numbered lengths on the line dropped from 8 in the end elevation. From all of the numbered intersections excepting 8′ and 9′ draw lines perpendicular to 1′ 8′ until they intersect the vertical line dropped from 7. Connect the various numbers in A as called for in the side elevation and then will these slant lines in A show the true lengths of similar lines in the side or end elevation.

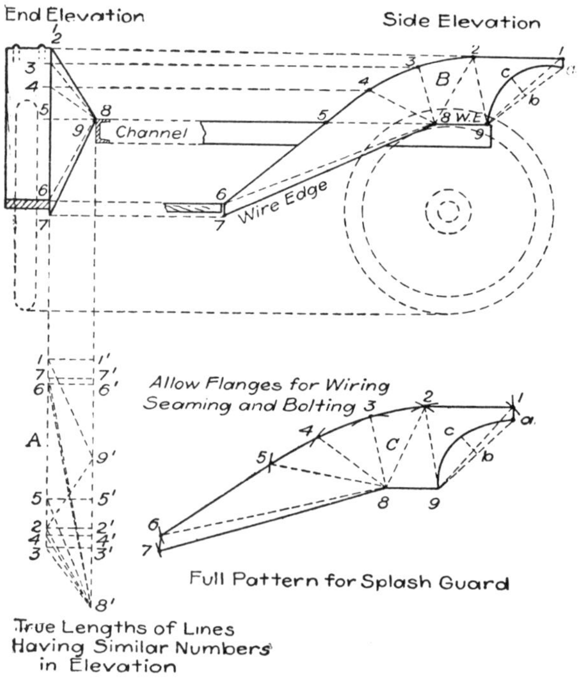

Fig. 134. Elevations, True Lengths, and Patterns for Mud Guard of the Type Shown in Fig. 133

For the pattern take the distance of 8 9 in the side elevation, which will be called B, and place it as shown by 8 9 in the pattern which will hereafter be called C. Now with 8′ 2 in A as radius and 8 in C as center, describe the arc 2, which intersect by an arc struck with 9 as center and 9′ 2 in A as radius. With 2 1 in B as radius and 2 in C as center draw the arc 1, which intersect by an arc struck with 9 as center and 9′ 1 in A as radius. With radii equal to 8′ 3, 8′ 4, 5, 6 and 7 in A and 8 in C as center draw the short arcs 3, 4, 5, 6 and 7. Next set the dividers equal to the various spaces 2 to 3 to 4 to 5 to 6 and 7 in the side elevation, and starting at the intersection 2 in the pattern C, step to similar numbered arcs as shown from 2 to 7 and draw a line from 7 to 8. Trace the outline from 1 to 7 in C as shown.

To reproduce the curve *a c* 9 in the side elevation on to the pattern, bisect *a* 9 in both B and C and obtain *b* in both the elevation and pattern respectively. At right angles to *a* 9 in the side elevation and from *b*, draw the line *b c* cutting the curve at *c*. Take this distance from *b* to *c* and set it off from *b* to *c* in the pattern C at right angles to *a* 9 and draw a graceful curve from *a* to *c* to 9. Then will 1 6 7 8 9 *c a* be the pattern for the splash guard of the type shown in Fig. 133, to which flanges must be allowed for wiring, seaming and bolting.

Pattern for Mud Guard Having A Double Curve

Fig. 135 shows another type of mud guard having a double curve, that is, the lower curve meets the running board at A while the upper one continues to B. Note that the lower part of the guard meets the running board filler at *a b* and from the inner edge of the running board *b* a bend takes place along *b c* meeting the channel frame at *c*. From *c* the metal runs at an angle to *d* then along the top of the channel *d e*, all as shown by the dotted line. By placing *c d* behind the fender line it allows the splash guard to have a flare at points *c* and *d* and not run flush with the surface of the fender, as was shown in the problem in Fig. 122 at 6 7 in the side elevation.

The method of working out the splash guard in Fig. 135 is shown in detail in Fig. 136. In this figure 1 6 7 8 to 10 shows the profile of the fender in the side elevation, 10 to 13 the joint line between the splash guard and the running board filler, shown from 10 to 13 in the end elevation. In the side elevation 10 to 14 shows the slight bend required between the rear of the running board and the lower side of the channel frame, 14 representing the apex, to which lines are drawn from points 7 to 10 and 10 to 13. A line is drawn from 14 to 15 parallel to 7 6 making the depth between 6 15 and 7 14 equal to the flare required on that part of the splash guard. The joint line between the bottom of the guard and top of the channel is shown by 15 to 16 while 16 *c a* shows the finish at the front of the guard and *a* 1 the width of the rim. Having completed the side elevation, draw the end elevation in its proper position as shown, making the projection between the fender and channel as much as required by the front wheels, or as much as is indicated in this case from 6 to 15. Divide the profile of the fender in the side elevation into any number of equal parts, as shown in this case from 1 to 7 to 10. Also divide the profile of the running board splash guard in the end elevation, as shown by 10, 11, 12 and 13. From these points project horizontal lines to the side elevation, cutting the joint line also from 10 to 13. From the various points 1 to 10 in the side elevation project horizontal lines to the end view, cutting the inner fender line also from 1 to 10. From points 13, 14, 15 and 16 in the side draw lines to the end elevation cutting the channel frame also from 13 to 16 as shown. From points 7 to 10 to 13 in both elevations draw lines to 14. In a similar manner in both views draw lines from 6 to 14, 4 and 5 to 15, 4, 3, 2 and 1 to 16. These lines in the side elevation are used as the bases of triangles, the altitudes of which are equal to the horizontal distances between similar numbers in the end elevation. From points 10 and 13 in the end elevation drop vertical lines as shown.

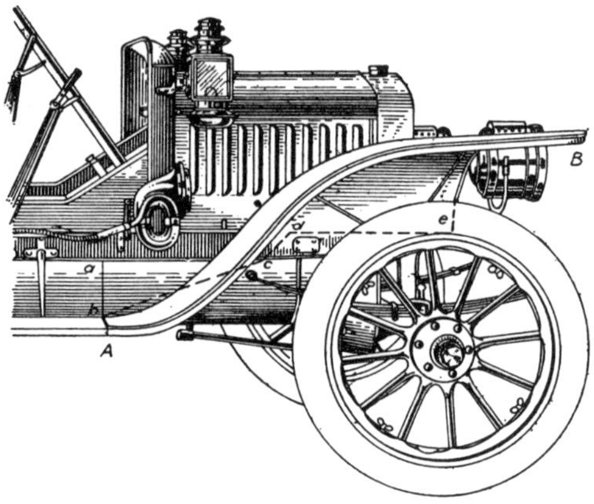

Fig. 135. Another Style of Mud Guard Having a Double Curve

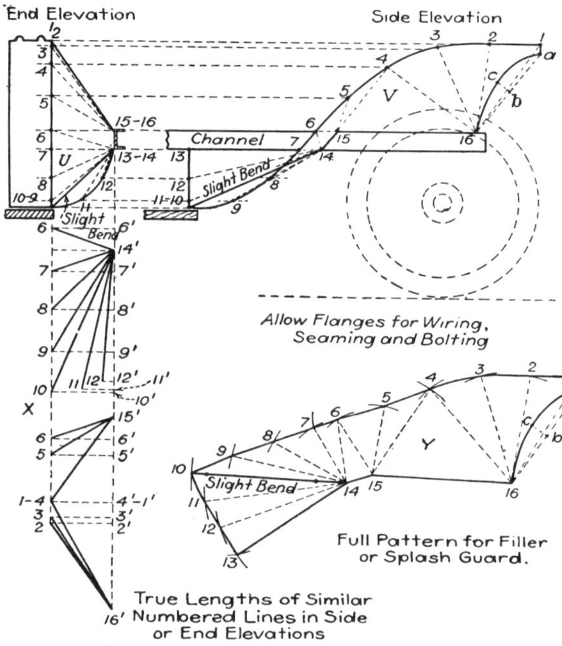

Fig. 136. Elevations, True Lengths and Pattern for Type of Fender Shown in Fig. 135

Now for example, to find the true lengths of the lines 10 14 and 4 16 in the side elevation proceed as follow: Take these distances 10 14 and 4 16 in the side elevation and place them on the vertical line dropped from 13 in the end elevation, as shown by 10′ 14′ and 4′ 16′. From 10′ and 4′ draw horizontal lines meeting the vertical line projected from 10 and 4 in the end view as shown by 10 and 4 respectively.

Draw the slant lines 10 14′ and 4 16′, which gives the true lengths of similar numbered lines in either elevation. In this manner the various distances of the lines in the side elevation are transferred to the diagram X, as shown by similar numbers on the vertical line 6′ 16′ and the true lengths found. It will be noticed that where the lines radiate from a given number in the side elevation they radiate from a similar number in the true lengths in X. For example 14′, 15′ and 16′ in X, are similar to 14, 15 and 16 in the side elevation.

The true lengths having been found, the pattern is now in order. Take the distance 15 16 in the side elevation or V and place it in diagram Y. With 15′ 4 in the true lengths or X as radius and 15 in Y as center draw the arc 4, which intersect by an arc struck with 16 as center and 16′ 4 in X as radius. With radii equal to 16′ 3, 2 and 1 in X and 16 in Y as center draw the arcs 3, 2 and 1. Set the dividers equal to the distances between 4 and 1 in V and starting from 4 in Y, step to arc 3, 2 and 1. Draw a line from 1 to 16. At right angles to 2 1 draw 1 *a* equal to 1 *a* in V and draw a line from *a* to 16 in both Y and V. Bisect these two lines and obtain *b*. From *b* in the side elevation draw the perpendicular *b c* cutting the curve at *c*. Take this distance *b c* and set it off from *b* in Y perpendicular to *a* 16 as shown by *b c*, and draw the curve *a c* 16. With radii equal to 15′ 5, and 15′ 6 in X and 15 in Y as center, draw the arcs 5 and 6. Set the dividers equal to 4 to 5 to 6 in the side elevation and starting from 4 in Y, step to arc 5 and 6 and draw a line from 6 to 15. Next with 15 14 in the side elevation as radius and 15 in Y as center draw the arc 14, which intersect by an arc struck with 6 as center and 6 14′ in X as center. With radii equal 14′ to 7, 8, 9, 10 11 and 12 in X, and 14 in Y as center draw the arcs 7, 8, 9, 10, 11 and 12. Set the dividers equal to the spaces between 6 and 10 in the side elevation and starting from 6 in Y step to arcs 7, 8, 9 and 10, and draw a line from 10 to 14. Again set the dividers equal to the divisions in the curve 10 13 in the end elevation U, and starting from 10 in Y step to arc 11 and 12, and from the intersection 12 strike the arc 13, which intersect by an arc struck from 14 as center and 14 13 in the side elevation as radius. Trace a line through points thus obtained in Y and then will 1 10 13 16 *a* be the full pattern for the splash guard shown in Fig. 135.

Pattern for Mud Guard with Fender curved both in Plan and Elevation, and intersected by Running Board Splash Guard

In Fig. 137 is shown the finished side and top views of a mud guard in which the fender is curved in plan as well as in the elevation and is intersected by the running board splash guard as shown in the elevation from A° to *d*′ and in the plan from *d* to A. In the elevation the fender meets the channel from A° to B° and then continues along the curve to C°. These same points of intersections and curve are shown in plan by A to B to C. The filler or splash guard is beaded as shown at X and meets the top of the channel frame along *a b* or *a′ b′* in elevation. The method of developing the patterns for the rim, fender, splash guard and running board splash guard is shown in detail in Fig. 138 and is developed as follows:

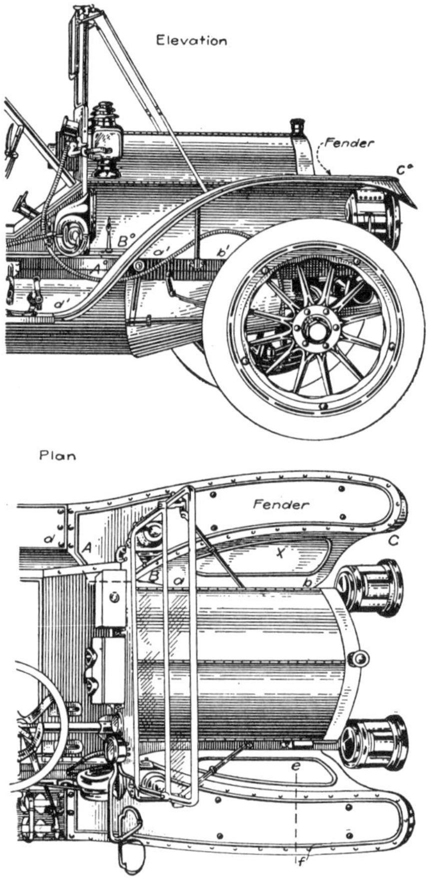

Fig. 137. Finished Side and Top Views of Mud Guard, in which the Fender is Curved in Elevation as well as in Plan

First draw the side elevation of the mud guard as shown by 5 11 16 Y X 11^v 5^v and in its proper position draw the sectional view of the running board splash guard as shown, which hooks around the bottom of the channel at 9 10 and flanges over the running board from 1 to 4. In its proper position below the elevation draw the plan view of the mud guard, obtaining the points of intersection against the channel at 8 11 from 8 11 in the elevation as shown by the dotted lines. Take a tracing of the sectional view in the elevation and place it in its proper position in plan as shown, and obtain the intersection between the outer edge of the fender and running board 5′, and from this point draw the desired curve 5′ 16′. Make 16′ 16 equal to the width of the top of the fender and draw a graceful curve from 16 to 11. The curves from 5′ to 16′ and from 16 to 11, may be drawn at pleasure, but the intersection between the running board splash guard and fender 5 to 8 in the plan must be projected as follows: Divide the curve of the running board splash guard in elevation as shown by the small figures 1 to 10, from which points draw horizontal lines until they intersect the curve of the fender in elevation as shown from 5 to 8. Establish an extra point between 5 and 6 in elevation and call it *a* which project back to the sectional view and obtain also *a*. Now take the distance from 5 to *a* and set it off from 5 to *a* in the sectional view in plan, and from the various divisions 5 to 8 draw horizontal lines into the plan, which intersect by vertical lines dropped from similar numbers in the elevation, thus obtaining the intersections 5 to 8 in the plan view. Trace a line through these points, which will give the joint line. Extend the lines just drawn until they intersect the outline of the fender in plan from 5′ to 8′.

To obtain the miter cut or pattern for the running board splash guard draw any line as A B, on which place the girth of the profile 1 to 10 in the sectional view, as shown by similar numbers. Through these small figures and at right

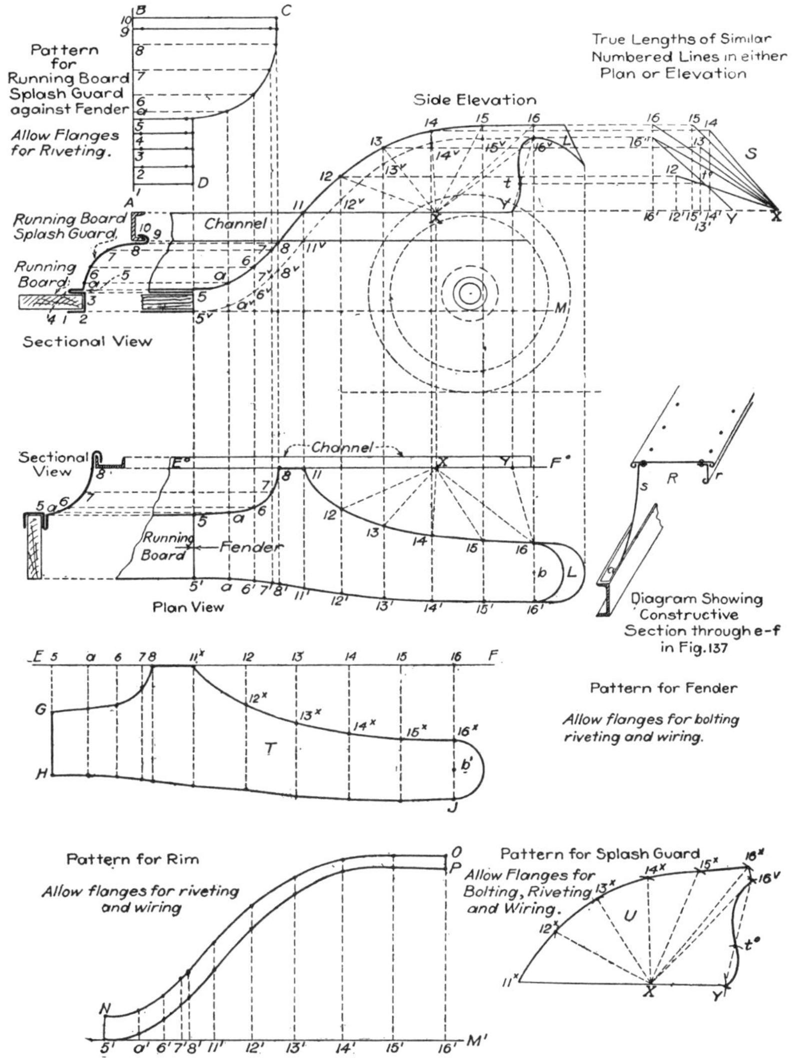

Fig. 138. Plan, Elevation, Sectional Views, True Lengths and Patterns for Mud Guard and Running Board Splash Guard Shown in Fig. 137

angles to A B draw lines as shown, which intersect by vertical lines erected from similar numbers in the elevation. A line traced through points thus obtained as shown from C to D will be the miter cut against the fender.

The pattern for the fender is next in order. Divide the rest of the curve from

8 to 16 in the side elevation into equal parts as shown, from which points drop vertical lines cutting the outline of the fender in plan from 11 to 16 and 11′ to 16′. Draw any line as E F on which place the girth of the profile of the fender from 5 to 16 in the side elevation, as shown by similar numbers on E F, through which perpendicular lines are drawn as shown. Now measuring from the line E° F° in plan take the various projections to intersections 5 to 8, 11 to 16 and 16′ to 5′, and place them on similar numbered lines in T, measuring in each instance from the line E F, and in that way obtain the points of intersections from G to 8, 11^x to 16^x and J to H. Trace a line through these points and then will the outline T show the full pattern for the fender, *b*′ showing the center from which the circular head is struck.

Draw the lower line of the rim in elevation as shown from 5^v to 16^v also the lip L. As the rim is curved in plan a pattern for it must be developed by drawing any line as 5′ M^1, on which is placed the girth of 5′ 16′ in plan as shown by similar numbers on 5′ M^1. From the small figures 5′ to 16′ and at right angles to 5′ M^1 erect lines indefinitely as shown. Measuring from the horizontal line 5^v M in the elevation, take the various heights to points 5, *a*, 6, 7 to 16 and place them on similar numbered lines, measuring in each instance from the line 5′ M^1. In a similar manner, again measuring from the line 5^v M in elevation take the various heights to points a^v, 6^v, 7^v, to 16^v and place them again on similar lines in the pattern for the rim, again measuring from the line 5′ M^1. Trace a line through points thus obtained and then will N O P 5′ be the pattern for the rim.

The pattern for the splash guard is next in order. Establish at pleasure the point X in elevation, so that X 11 will be similar to X 16 and draw lines from 12, 13, 14, 15, 16 and 16^v to X. At pleasure draw the ogee curve 16^v *t* Y and draw a line from 16^v to Y. In a similar manner connect similar numbers to X and Y in plan, obtaining points X and Y in plan from X and Y in elevation. These lines in plan will then represent the bases of triangles, the altitudes of which are equal to the vertical heights between similar numbers in elevation and are shown in diagram S, where the distances along the base line 16′ X are obtained from the lines in plan and the heights in S are obtained from the elevation as shown. Thus X 16 in S shows the true length of X 16 in elevation or plan. The pattern for the splash guard is shown in U in which X 11^x is equal to X 11 in elevation. Now with radii equal to X 12, 13, 14, 15 and 16 in the diagram of true lengths in S and X in the pattern U as center, draw the arcs 12^x to 16^x. Set the dividers equal to the spaces between 12 and 16 in the side elevation and starting from 11^x in U step to arc 12^x, etc., to 16^x. With 16 16^v in the side elevation as radius, and 16^x in

U as center draw the arc 16^v, which intersect by an arc struck with X as center, and X 16″ in diagram S as radius. With X Y in the side elevation as radius and X in U as center draw the arc Y, which intersect by an arc struck from 16^v as center and 16″ Y in S as radius. Where the ogee 16^v *t* Y in the side elevation crosses the line 16^v Y at *t* project this point to the true length in S by drawing a horizontal line from *t* in elevation until it crosses the true length 16″ Y in S at t°. Take the distance from Y to t° and set it off from Y to t° in the pattern U and trace a free hand ogee through points 16^v, t° and Y. Trace a curved line from 16^x to 11^x, which completes the pattern for the splash guard.

Diagram R shows a constructive section through *e f* in Fig. 137, and shows in Fig. 138 in R how the bottom flange of the filler *a* sets on the top of the channel flange of the chassis and how it is riveted to the fender at the top. As will be seen, the fender is beaded on both sides, the rim being riveted to the outer edge of the fender. In the previous construction of the mud guard it will be remembered that the filler *s* was seamed and the rim *r* locked. The lip L in elevation is developed as explained in Fig. 132.

Pattern for a Flaring Fender

Fig. 139 shows a finished view of a flaring fender without a rim. Note the outline of the vertical splash guard from *a* to *b* to *c*. Sometimes this same flaring fender has a rim along A B. The method of developing the flare is similar to cutting the frustum of a right cone and is shown in Fig. 140, in which 1 9 B A shows the elevation of the fender struck from the

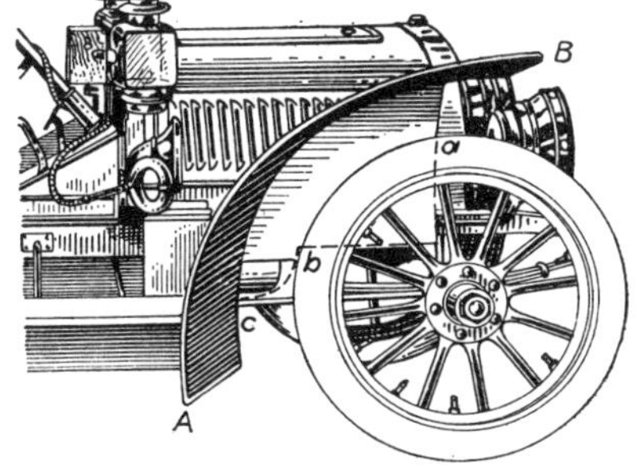

Fig. 139. Finished View of a Flaring Fender without Rim

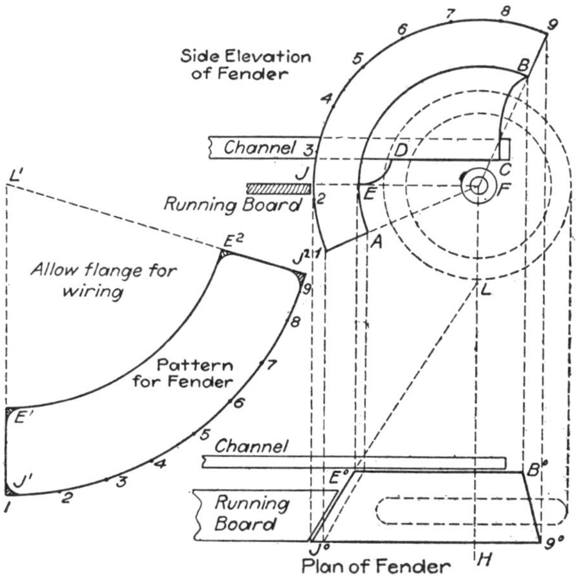

Fig. 140. Plan, Elevation and Pattern for Flaring Fender Shown in Fig. 139

center of the axle F. The outline of the vertical splash guard, which having no flare, is used as the pattern is shown by B C D E. Directly below the elevation draw the plan as shown by J° E° B° 9°, which shows a horizontal section through J F in elevation. From the center F draw the center line F H, which intersect at L by extending the flaring line J° E°. Divide the outer curve 1 9 in elevation into equal spaces as shown. In plan, L E° and L J° then become the radii with which to develop the pattern and the spaces from 1 to 9 in elevation give the amount of material required. Using the radii just mentioned and L^1 as center draw the arcs E^1 E^2 and J^1 J^2. From J^1 draw a radial line to L^1 intersecting the inner arc at E^1. Starting from J^1 step off the girth of 1 to 9 in elevation and from the point 9 in the pattern draw a line to L^1 cutting the inner arc at E^2. The pattern for the flaring fender as called for in Fig. 139 is then represented in Fig. 140 by E^1 J^1 J^2 E^2. If desired the corners of the pattern are rounded as shown by the shaded portion.

Pattern for Flaring Twisted Fender

Fig. 141 shows another type of flaring fender which is flaring at the bottom and gradually twists so that it becomes level at the top. The vertical splash guard is indicated by *a b c d*. A twisted fender of this kind requires triangulation in its development and is shown developed in Fig. 142. In this figure 10 6 *a* in the side elevation shows the inner curve of the fender and *a* 6 3 1 the outer curve of the fender. The inner curve shows a section on the line 10 6 in plan and the outer curve a section on the line 1 5 in plan. A line drawn from 1 to 10 in plan gives the flare at the bottom of the fender, while the top of the fender is level as indicated by *a* 6 in the side elevation. The pattern for the vertical splash guard shown by *a b c d* 10 8 6 *a* can be traced directly from the detail drawing, and edges allowed for wiring, seaming and bolting.

The pattern for the fender is obtained as follows: Divide both curves in the side elevation into equal parts as shown from 1 to 5 on the outer curve and 6 to 10 on the inner curve. From these points project lines to the plan intersecting the outer and inner curves as shown respectively from 1 to 5 and 6 to 10. Connect lines in the plan and elevation from 2 to 9, 3 to 8, 4 to 7 and 5 to 6 and diagonal lines from 10 to 2, 9 to 3, 8 to 4, and 7 to 5. The lines in plan then represent the bases of triangles which will be constructed in B, the altitudes of which will be equal to the vertical distances between similar numbered points in the side elevation, and are obtained as follows: Extend the line 10 1 in the side elevation as shown by 10 10, on which place the length of the lines in plan as

2 10, 3 9, 4 8 and 5 7, as shown by similar numbers on the line 10 10 in B. From these points at right angles to 10 10 erect perpendicular lines, which intersect by lines drawn from similar numbers in the side elevation parallel to 10 10. Connect these points of intersections as shown by heavy lines which will be the true lengths of the similar numbered lines in plan or elevation. As the lines in elevation from 2 to 9, 3 to 8, 4 to 7 and 5 to 6 lie on horizontal planes similar numbered lines in plan will then show their true lengths. In plan A shows the semi-circular end of the fender. The pattern is now developed by taking a reproduction of A 5 6 in plan and placing it as shown by A° 5 6 in the pattern C. Now with 6 7 in the side elevation as radius, and 6 in C as center draw the arc 7′, which intersect by an arc struck with 5 as center and 5′ 7′ in the true lengths in B as radius. With 5 4 in the side elevation as radius and 5 in C as center draw the arc 4′, which intersect by an arc struck with 7′ as center and 7 4 in plan as radius. Proceed in this manner, using alternately, first the divisions in the curve 6 10 in the side elevation as

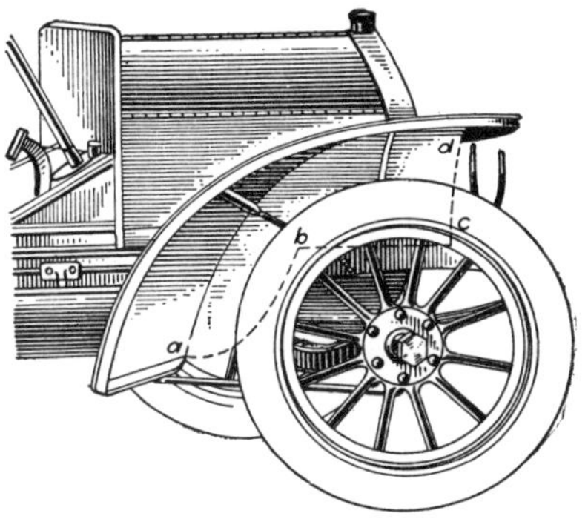

Fig. 141. Another Type of Flaring Twisted Fender, Requiring Triangulation in its Development

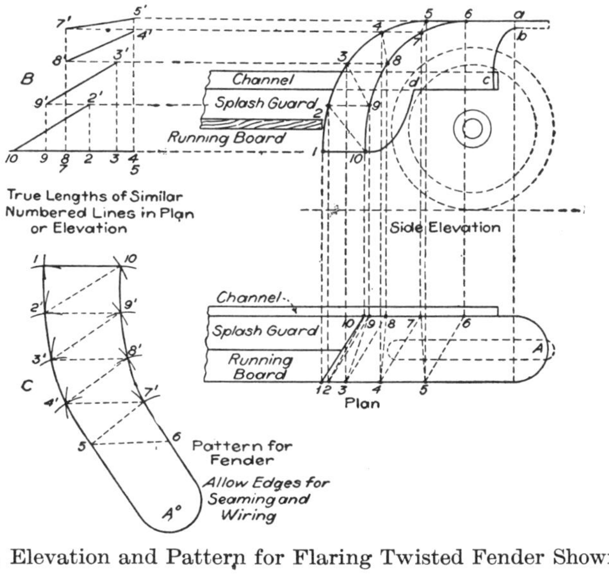

Fig. 142. Plan, Elevation and Pattern for Flaring Twisted Fender Shown in Fig. 141

radius, then the proper length in B; the proper division in the curve 1 5 in elevation as radius, then the proper length in plan, until the line 1 10 in pattern C has been obtained. Trace a line as shown by A° 1 10 which completes the pattern for the twisted flaring fenders.

Hind Wheel Fenders

Fig. 143 shows a finished view of a hind wheel fender connected to the running board at A, with a splash guard as indicated by *a b*. The method of developing these hind wheel fenders are similar to the front wheel fenders already described.

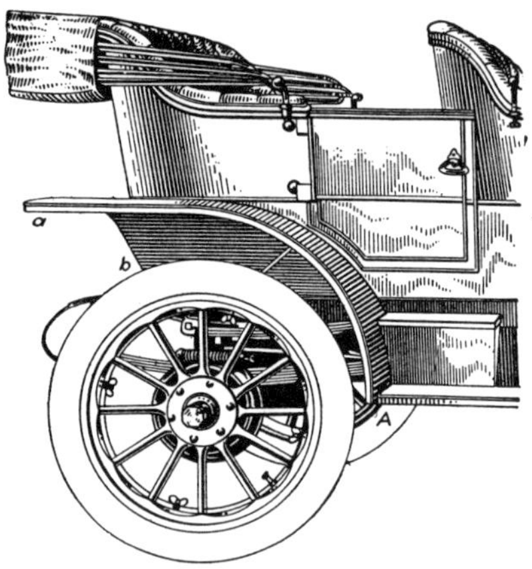

Fig. 143. Mud Guard on Hind Wheel

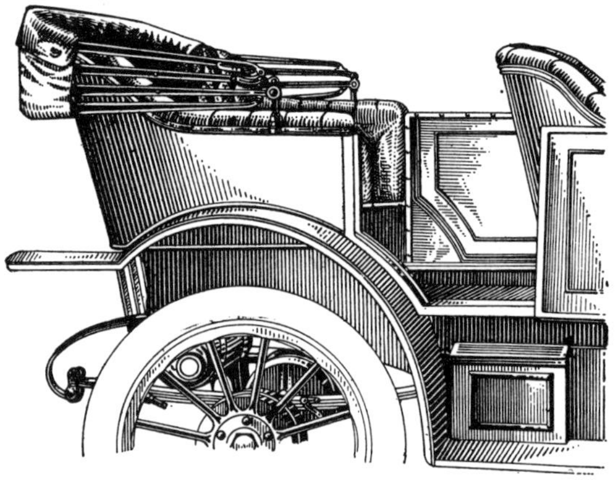

Fig. 144. Rear Fender Extending into the Wheel Housing in the Tonneau Seat Panel

On some of the automobiles the fenders are built into the wheel housing in the tonneau seat panel as shown in Fig. 144, the rims being seamed or riveted to the fender as already described.

Separate Dash Hoods

On some automobiles the dash hoods are made separate, and then set on to the body, making a finish with a brass moulding or fastening by means of angles on the inside of the body. In this part, four different types of separate dash hoods will be taken up. The first type is shown in Fig. 145 and shows a dash hood fastened separately to the dashboard *a b* and body *b c*, making a finish by means of the semi-elliptical brass moulding as shown. The method of developing this type of dash hood is very simple and is shown in Fig. 146 in which 1 1 5 6 6 shows the side elevation of the dash hood, an outline of which is shown in the front elevation. Divide the semi-outline from 1 to 6 in the

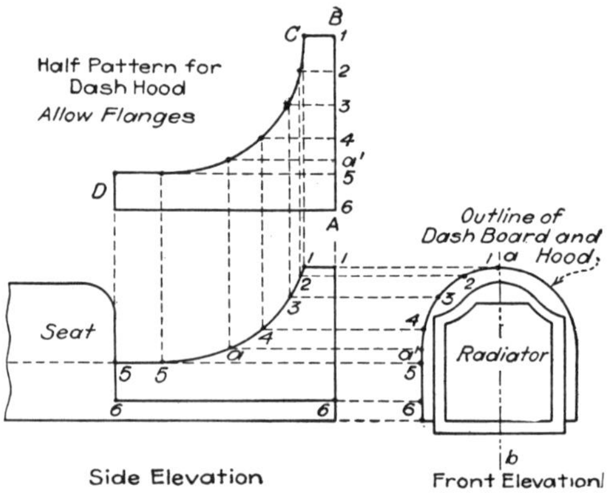

Fig. 146. Pattern for Separate Dash Hood Shown in Fig. 145

front elevation into equal spaces as shown, from which horizontal lines are projected to the side elevation intersecting the outline as shown. As the space between 4 and 5 in the side is very wide, establish an extra point as *a*, which project to the front elevation, thus obtaining *a'*. Now extend the line 6 1 in the side elevation as A B, on which place the girth of the semi-section of the dashboard in the front elevation, as shown by similar numbers on A B. Through these points and

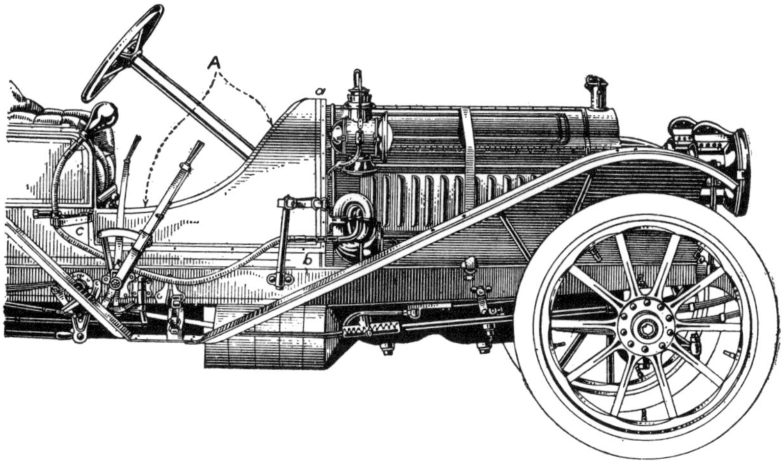

Fig. 145. View of Separate Dash Hood Forming Part of Body

at right angles to A B, draw lines as shown, which intersect by lines drawn parallel to B A, from similar numbered intersections in the side elevation. A line traced through points thus obtained as shown by A B C D will be the half pattern for the dash hood, to which a flange must be allowed along A D for securing to the body of the car.

Another Type of Separate Dash Hood Requiring Triangulation in its Development

In Fig. 147 is shown another type of separate dash hood with a flare along the top and sides. The dash hood is fastened to the dashboard along *a b* and to the body along *b c d*. The method of drawing the detail and developing the pattern is shown in Fig. 148, in which 7 8 9 *a b* shows the side elevation of the front part of the body over which the dash hood 1 7 8 9 13 sets. In line with the side elevation as shown, draw the front elevation of the radiator, dashboard and hood, the profile 8 13 8^x showing the section on the line 13 *a* in the side elevation, the profile 8 1 8^x in the front elevation the vertical section through 1 5 8 in the side elevation and the profile 7 1 7^x in the front elevation, the vertical section through 1 7 in the side elevation.

The first step is to divide the upper outline of the hood in the front elevation into equal spaces as shown from 1 7, and from these points draw horizontal lines into the side elevation cutting the proper outline as shown from 1 to 7. As the division between 5 and 7 is very wide, the point 6 has been established and is projected to the proper outline in the front elevation also at 6. From the points 9 and 8 in the side elevation draw lines to 5 and from 6 and 7 draw lines to 8. Connect lines in a similar manner in the front elevation. As there remain four spaces between 5 and 1 in both elevations, divide the lower outline of the dash hood in the front elevation also into four spaces as shown, and project the points

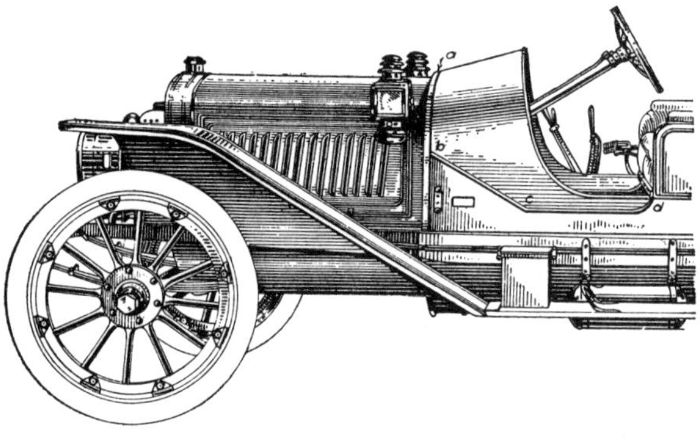

Fig. 147. Another Type of Separate Dash Hood Forming Part of Body

horizontally to the side view, as shown by similar numbers. Connect lines in both views as shown. Then will the lines in the front elevation show the bases of the triangles the altitudes of which are equal to the horizontal distances between similar numbered points in the side elevation and are obtained as follows: Extend the line *a* 13 in the side elevation as D E, on which place the various distances obtained from the front elevation as shown by similar numbers between 6′ and 13′ on D E. From the points 2′ to 8′ inclusive and at right angles to D E draw lines as shown, which intersect by lines drawn parallel to D E from similar numbered intersections in the side elevation. Connect points as shown by the heavy line, which gives the true lengths of similar numbered lines in either elevation.

The next step is to find the developed section or true edge line along 1 7 in the side elevation. This is obtained by taking this girth from 1 to 7 and placing it on the center line A 1 extended in the front elevation as B C. Through the small figures 2 to 7 and at right angles to B C draw lines, which intersect by lines drawn parallel to B C from similar numbers in the front elevation, thus obtaining the points of intersections 2^v to 7^v. A line traced through these points will give the true edge line along 1 7 in the side elevation.

To save time in describing the pattern, the developed section will be called U; the true lengths V; and the pattern W. First take the distance 1 13 in the side elevation which shows its true length, and place it as shown by 1 13 in W and with radius equal to 1 2^v in U and 1 in W as center, describe the arc 2, which

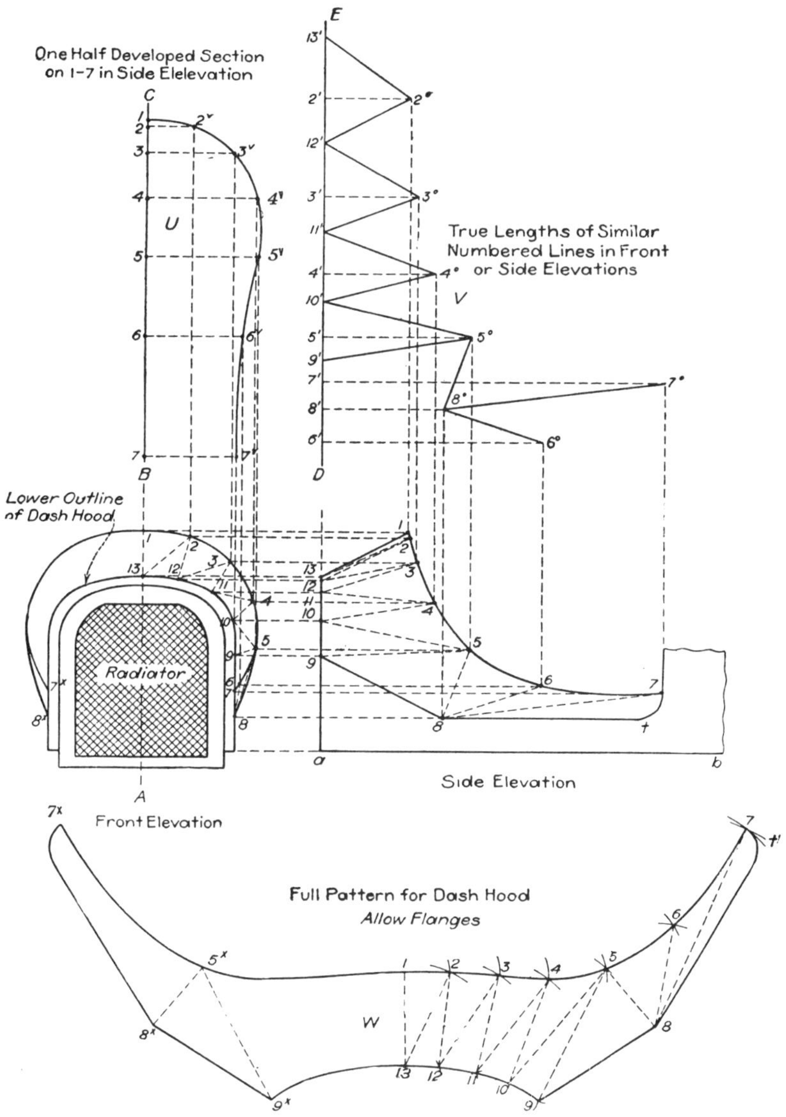

Fig. 148. Front and Side Elevations, One Half Developed Section, True Lengths and Pattern for Separate Dash Hood Shown in Fig. 147

intersect by an arc struck from 13 as center and 13′ 2° in V as radius. With 13 12 in the front elevation as radius and 13 in W as center, describe the arc 12, which intersect by another arc struck from 2 as center and 2° 12′ in V as radius. Proceed in this manner until the line 5 9 in the pattern W has been obtained.

Then using 9 8 in the side elevation as radius and 9 in W as center describe the arc 8, which intersect by an arc struck from 5 as center and 5° 8° in V as radius. Then with radii equal to 8° 6° and 8° 7° in V and 8 in W as center, describe the arcs 6 and 7. Set the dividers equal to the divisions 5^v, 6^v, and 6^v 7^v in U and starting from 5 in W, step to arc 6 then to 7 and draw a line from 7 to 8. Take a tracing of 8 *t* 7 in the side elevation and place it as shown by 8 *t* 7 in W. Trace a line through the various intersections in W and 1 7 *t* 8 9 13 will be the half pattern for the separate dash hood. For the complete pattern trace the half pattern opposite the line 1 13 as shown by 1 5^x 7^x 8^x 9^x 13. Flanges must be allowed for fastening to body.

Pattern for Separate Dash Hood Joining Door Frame

Fig. 149 shows another type of dash hood joining the dashboard, body and door frame at *a b*, *b c* and *c d* respectively, the dashboard having a different profile from that shown in the previous problem, and is shown detailed and developed in Fig. 150 in which the lower part of the body is shown by *a b* in the side elevation. The door is also shown. Adjoining the door frame and setting on the body is shown the side view of the dash hood 1 6 *a b* 11 1. In its proper position in line with the side elevation draw the front elevation of the dash hood. It makes no difference what shape the dashboard or hood may have, the principle shown is applicable to any shape. In this case the outline *r s t* 11 7 *r* in the front elevation shows the true profile of the dashboard on the line 11 *r*′ in the side elevation and the outline 6 1 6′ in the front elevation shows the vertical section on the line 1 6 in the side elevation.

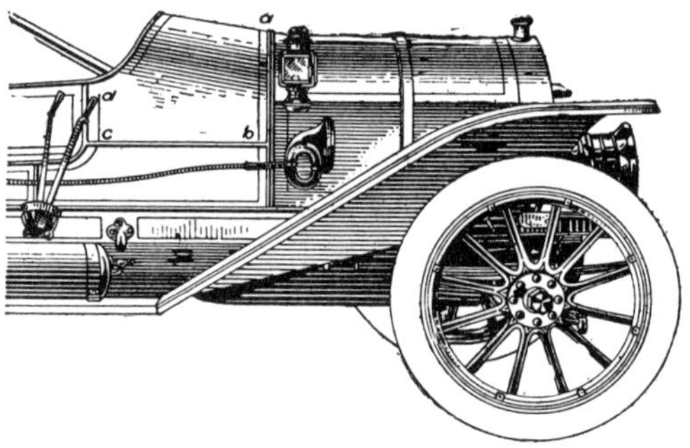

Fig. 149. Another Type of Separate Dash Hood Joining Door Frame

The first step is to divide the outline of the lower and upper part of the hood in the front elevation into equal parts as shown respectively from 7 to 11 and 1 to 6. It will be noticed that the divisions between 4 and 6 are smaller so as to obtain an extra division between 4 and 6 in the side elevation. From the divisions 1 to 6 and 7 to 11 in the front elevation draw horizontal lines to the side elevation, intersecting respectively the upper curve of the dash hood from 1 to 6 and the lower part of the hood on the line of the dashboard from 7 to 11. Connect the various points in the front elevation by lines drawn from 6 5 and 4 to 8, then from 4 to

9 to 3 to 10 to 2 to 11 and in a similar manner connect similar numbers in the side elevation.

The true length of these lines are obtained by taking the various distances of the lines in the front elevation and placing them on the line r' 11 in the side elevation extended as D E, as shown by similar numbers. From these small figures

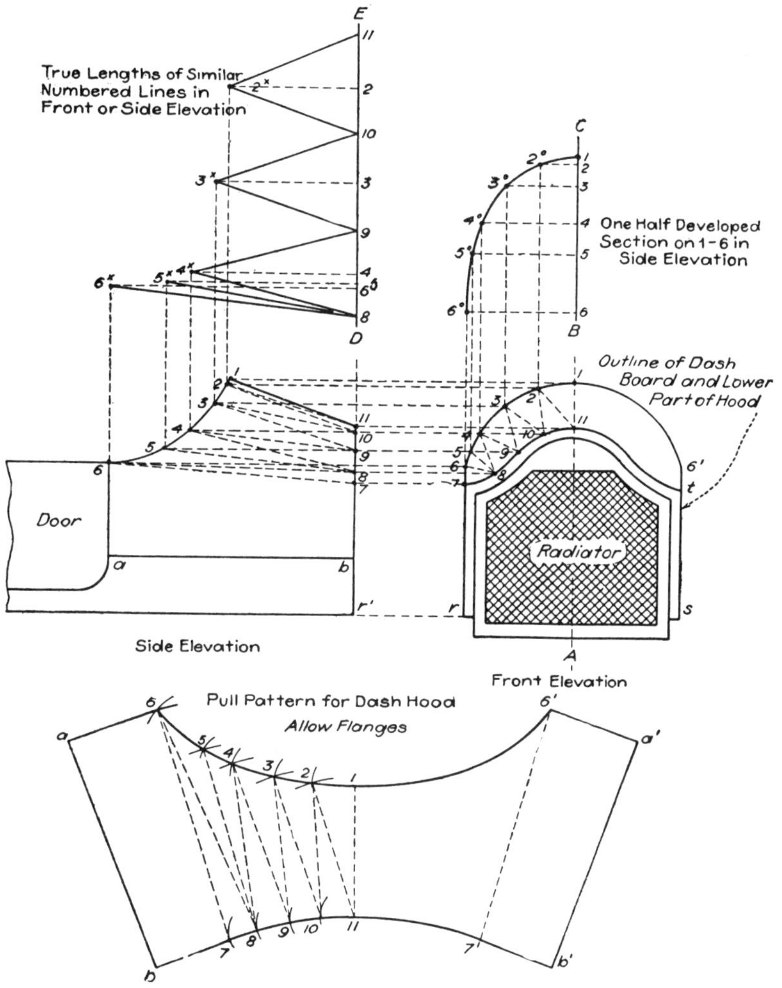

Fig. 150. Elevations, True Lengths, Half Developed Section and Full Pattern for the Dash Hood Shown in Fig. 149

and at right angles to D E, draw lines, which intersect by lines drawn parallel to D E from similar numbered intersections on the curve 1 6 in the side elevation. The solid lines shown in the true lengths then represent the true lengths of similar numbered lines in either elevation.

The true edge line of the curve 1 6 in the side elevation is obtained by taking the girth of this curve 1 6 and placing it on the line A B extended in the front elevation as shown by similar numbers on B C. Through these small figures and at right angles to B C draw lines which intersect by lines drawn parallel to B C from similar numbers in the front elevation, thus obtaining the points of intersections 2° to 6°, through which a line is traced from 1 to 6° and represents the one-half developed section on 1 6 in the side elevation.

The pattern is now in order and is obtained by taking the distance 1 11 in the side elevation and placing it as shown by 1 11 in the pattern. With 1 2° in the developed section as radius and 1 in the pattern as center describe the arc 2, which intersect by an arc struck from 11 as center and 11 2^x in the true lengths as radius. Using 11 10 in the front elevation as radius and 11 in the pattern as center, draw the arc 10, which intersect by an arc struck from 2 as center and 2^x 10 in the true lengths as radius. Proceed in this manner until the line 6 7 in the pattern has been obtained, which is equal in length to 6 7 in the side elevation. The divisions from 1 to 6 in the pattern are taken from 1 to 6° in the developed section and the division from 11 to 7 in the pattern are taken from 11 to 7 in the front elevation, while the dotted lines in the pattern are obtained from similar numbered slant lines in the true lengths. A tracing is now taken of *a b* 7 6 in the side elevation and placed as shown by *a b* 7 6 in the pattern. 1 6 *a b* 11 then represents the half pattern of the dash hood. If the pattern is desired in one, simply trace the half opposite the line 1 11, as shown by 1 $6'$ a' b' $7'$ 11. The necessary flanges must be allowed as the pattern is shown net.

Pattern for Separate Dash Hood on a Torpedo Model

Fig. 151 shows another type of a separate dash hood joining the front part of the body and door frame, the hood being cut on a horizontal line as shown at *a*.

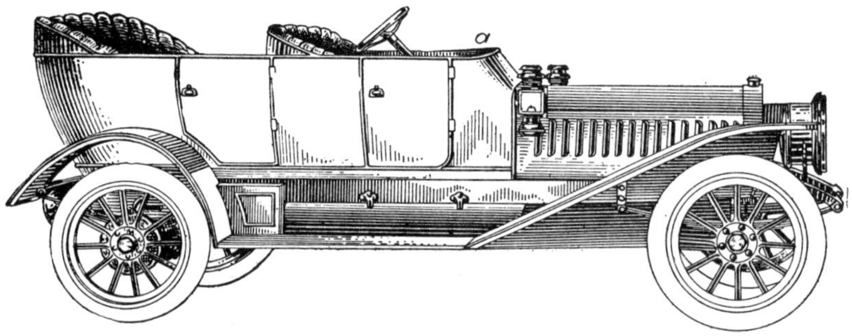

Fig. 151. Another Type of Separate Dash Hood Meeting Door in a Torpedo Model

When the profile or shape of the dashboard is known, and the profile of the upper opening of the hood at *a* is given, the pattern must then be developed by triangulation as shown in detail in Fig. 152, in which the side elevation of the door, body and dash hood is indicated by *c* 10 1 A, Y, and the front elevation directly in line with the same as shown. Above the side elevation, draw the top view of the hood as shown, making the width from 6 to *r* equal to the width from *a* to *b* in the front elevation. Through the plan view draw the center line F G and through the front elevation the center line D E. Divide the half profile of the dashboard or lower end of the hood into equal parts, as shown from 1 to 5 or four

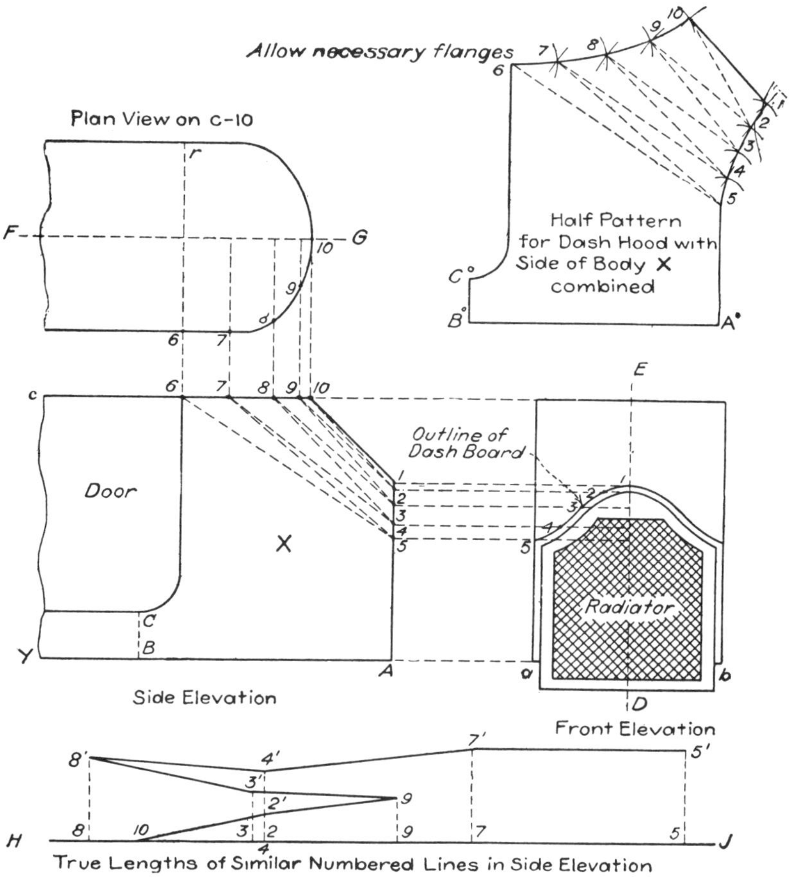

Fig. 152. Elevation, Plan, True Lengths and Half Pattern for Dash Hood in a Torpedo Model Shown in Fig. 151

spaces and divide the half profile of the plan also into four spaces, as shown from 6 to 10. At right angles to F G and through the small figures 6, 7, 8 and 9, draw lines meeting the line F G also the line *c* 10 in elevation as shown by similar numbers. In a similar manner and at right angles to E D in the front elevation, through points 1 2 3 4 and 5 draw lines cutting the center line E D and the lower line of the dash hood in the side elevation from 1 to 5. Connect the various points in the elevation as shown. These lines then represent the base lines of sections which will be constructed, the altitudes of which are equal to the various distances in the half plan and half front elevation.

As 1 10 and 5 6 in the side elevation show their true lengths, then take the various other lengths of the lines shown in the side elevation and place them as shown by similar numbers on the line H J. For example, to find the true length of 7 5 in the side elevation, take this length, and place it as shown by 7 5 on the line H J. At right angles to H J and from points 7 and 5, erect the lines 7 7′ and 5 5′, equal respectively to the distance measured from the line F G to the point 7 in the plan view, and the distance measured from the line D E in the front elevation to the point 5. A line drawn from 7′ to 5′ in the true lengths will give the true length of 7 5 in the side elevation. In this manner all of the true lengths of the various lines in the side elevation are obtained, as shown by similar numbers in the diagram of true lengths.

In developing the pattern a seam is made along 1 10 in the side elevation and the pattern is developed as follows: Take a tracing of A B C 6 5 in the side elevation and place it as shown by A° B° C° 6 5 in the pattern and with 6 7 in the plan as radius and 6 in the pattern as center describe the arc 7, which intersect by an arc struck from 5 as center and 5′ 7′ in the true lengths as radius. With 5 4 in the front elevation as radius and 5 in the pattern as center describe the arc 4, which intersect by an arc struck from 7 as center and 7′ 4′ in the true lengths as radius. Proceed in this manner, using alternately, first the proper division in plan, then the proper length in the true lengths; the proper division in the front elevation, then again the proper length in the true lengths, until the line 10 1 in the pattern has been obtained, which is equal to 10 1 in the side elevation. Trace a line through points thus obtained and A° B° 6 10 1 A° will be the half pattern for the dash hood and part of body combined.

Flanges must be allowed for riveting, bolting and the like.

CONSTRUCTING TANKS AND MISCELLANEOUS PARTS

Gasoline Tanks

The gasoline storage tanks are usually made from Nos. 22 or 24 gauge galvanized sheet steel or 16 oz. cold rolled copper. On the touring cars the tank is usually made oblong in shape and placed under the front seat or sometimes the storage tank is made of copper, round or elliptical in section and placed on the outside of the rear of the car. The way the rectangular tank is constructed is shown in Fig. 153. It will be noticed that the body is made in one piece, doubled seamed at the corner A, and the locked joint thoroughly sweated with solder, holes of course being punched in the sides of the tank to receive the brass fillers and stoppers, which are sweated with solder. Before the head B is placed at the end of each tank, partitions shown by *a b c d* are soldered in position, with openings in each corner at *a b c* and *d*, also at *e*. These partitions are made of number 20 gauge galvanized iron or of 20 oz. sheet copper, hard rolled, and act

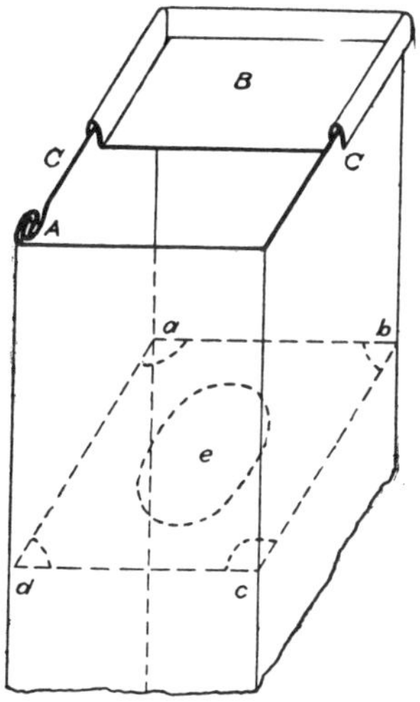

Fig. 153. Constructing the Galvanized Iron Gasoline Tank

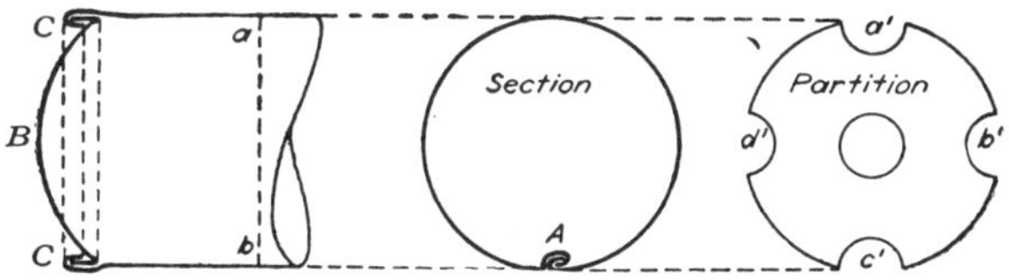

Fig. 154. Constructing the Copper Gasoline Tank

as stiffeners inside of the tank, also as splashers, that is, to keep the gasoline from splashing inside of the tank. Usually two splashers are placed in each tank. The head B is now locked at C and C and thoroughly sweated with solder. By referring to the view in Fig. 126, a group of finished galvanized iron tanks are shown, X X showing the bodies with brass cups soldered in position, while y y show the heads.

Fig. 154 shows the constructive view of the round or elliptical gasoline tanks usually made from sheet brass or copper. As will be noticed the body is locked at A and soldered. The head B is raised and flanged outward as shown, over which the ends of the body are seamed at C and C and well soldered. Splashers are soldered on the inside at *a b*, the shape of the splasher or partition being shown by *a' b' c' d'*.

The Oil Tank

The oil tank is usually made of IXX tin with the heads seamed in same, the tank being oblong with rounded corners. This tank has a filler plug at the top, in which the oil is poured, and a discharge at the bottom which supplies oil to the motor. A glass tube is placed in front so that the amount of oil in the tank may be seen. This tank is usually placed in front of the dashboard under the motor hood, with the glass tube showing on the outside of the dashboard toward the rider.

Fly Wheel Hood

The fly wheel hood or cover is used, as its name indicates, to cover the fly wheel at the bottom of the car and is fastened to the car by bolting to the bottom of the radiator and channel frame. It is made of No. 24 gauge black sheet steel in shape as required by the different constructions with all joints double seamed.

Mud Apron or Pan

The mud apron or pan is made of No. 24 gauge black sheet steel and is shown in position in Fig. 1. It is used and made to protect the lower part of the motor from dust, mud and flying gravel. Lock nuts are used when bolting to the channel frame, which avoids the loosening of same by the vibration of the car. This mud pan is usually made in two halves, with hooks riveted on each side, and these hooks secured by spring clips accessible from the sides.

The Muffler

The muffler is used to muffle or deaden the noise of the escaping gases. It is a cylinder shaped tube made up of No. 24 gauge black sheet steel, inside of which are placed iron pipes, through which the gases escape. The joints are all seamed and riveted.

SHEET METAL PART OF THE EQUIPMENT

Types of Funnels Used

Each automobile carries a full equipment of tools, wrenches, and the like, and in this equipment are sometimes added the various funnels for filling oil and gasoline. These funnels are of such different shapes from those usually made up in the tin shop, that four of the different shapes have been selected and their method of development will be described. These funnels are sometimes made up in the automobile shops, although frequently they are made in the sheet metal shops anywhere, which sell direct to the owners of the machine as they pass through the country towns on their tours, or are sold by automobile supply stores in the larger cities. They are usually made of 16 oz. cold rolled copper or of sheet brass. Three shapes of funnels will be taken up and one measure with funnel combined. The measures are made up in any size, from ½ pint to 1 gallon.

Pattern for Center Funnel Oval to Round

In Fig. 155 is shown the method of obtaining the patterns for the oval funnel and round spout and $1'$ $1''$ 9^v 9 shows the elevation of the funnel and 9 9^v 10^v 10 the elevation of the spout. The shape of the funnel at the top is shown by the plan of the top, while the half section on 9 9^v is shown by 9 6 9^v and is struck from the center $6'$. Through the center of the elevation and plan draw the center line A B and lengthwise through the plan the center line C D. Divide one half of the semi-circle into equal spaces as shown by the small figures 6, 7, 8 and 9. Extend the side of the spout 9 10 until it meets the center line A B at E. Then using E as center and radii equal to E 10 and E 9, draw the arcs 10^v 10^a and 9^v 9^a respectively. On the outer arc establish any point as $9''$, from which draw a line to the center E intersecting the inner arc at $10''$. Take four times the girth of the quarter circle 9 to 6 in the elevation and place it on the outer arc 9^v 9^a, starting from the points $9''$, thus obtaining points from $9''$ to 6 to 9 to 6 to 9^a. From 9^a draw a line to the center E, cutting the inner arc at 10^a and $9''$ 9^a 10^a $10''$ then shows the pattern for the spout without any laps.

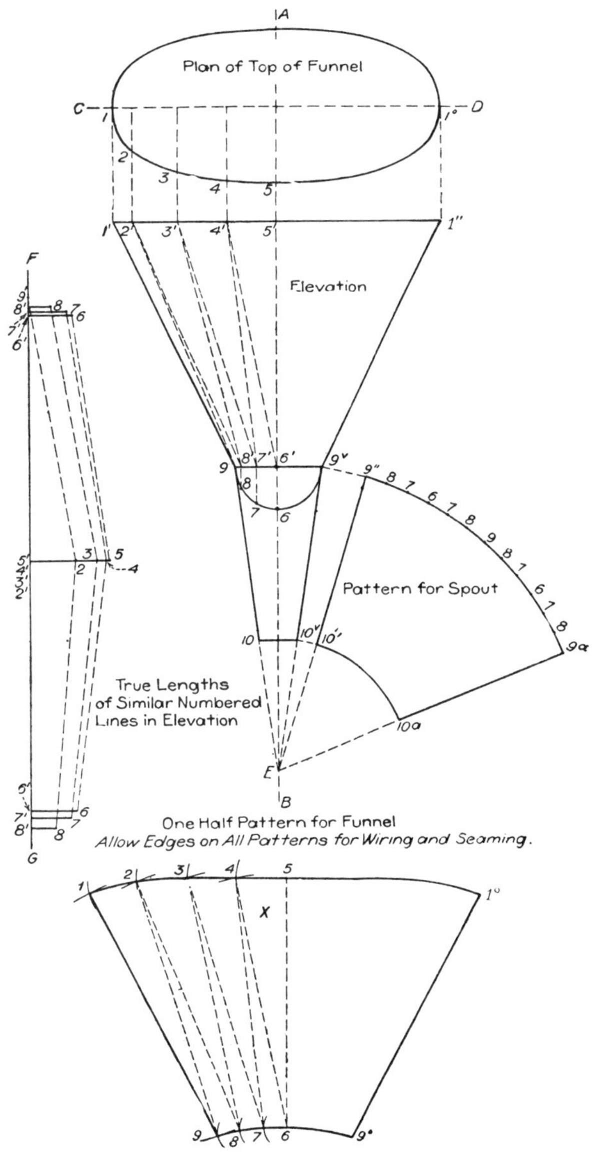

Fig. 155. Obtaining Pattern for an Oval to Round Funnel

As the funnel is oval at the top and round at its base, the pattern will be developed by triangulation as follows: Divide the one quarter oval plan into equal spaces as shown from 1 to 5, from which points and at right angles to C D draw lines intersecting the line C D, as well as the top line 1′ 1″ in elevation as shown by the small figures 1′ to 5′. As there are four spaces in the quarter oval, and but three spaces in the quarter circle, connect 5′ and 4′ to 6′ and the rest of the points by lines as shown. These lines then represent the bases of sections which will be constructed, the altitudes of which are equal to the various perpendicular heights in the quarter oval and quarter circle. For example: to find the true length of the line 3′ 7′ in elevation, take this distance and place it on the vertical line F G as shown from 3′ to 7′. From these two points and at right angles to F G draw the line 3′ 3 and 7′ 7 equal respectively to the distances measured from the line C D in the quarter oval to the point 3, and from the line 9 9ᵛ in the semi-circle to the point 7. A line drawn from 3 to 7 in the diagram of true lengths will give the true length of the line 3′ 7′ in the elevation.

Assuming that the funnel is to be made up in two parts with a seam at 1 and 1° in the plan, start the half pattern by taking the distance 5 6 in the true lengths and placing it as shown by 5 6 in the pattern X. Now using 5 4 in the half plan as radius and 5 in X as center describe the arc 4, which intersect by an arc struck from 6 as center and the true length 6 4 as radius. Then with 6 7 in the semi-circle in elevation as radius and 6 in X as center describe the arc 7, which intersect by an arc struck from 4 as center and 4 7 in the true lengths as radius. Proceed in this manner, using alternately first the divisions in the oval plan, then the proper true length,—the divisions in the semi-circle in the elevation, then again the proper true length, until the line 1 9 in the pattern has been obtained, from 1′ 9 in the elevation. Trace this quarter pattern 1 5 6 9 opposite the line 5 6 as shown by 1° 9°. Then 1 1° 9° 9 will be the half pattern for the funnel to which edges must be allowed for seaming and wiring.

Pattern for Funnel, Perpendicular at One End

In Fig. 156 is shown how the patterns are developed for a funnel, the top of which is oval and the spout round, and when one end of the funnel is perpendicular. First draw the side elevation of the funnel as shown, making the pitch as desired. Draw the oval section of the top of the funnel as shown above the elevation, representing the true section on the line 1′ 11′ in the side elevation. On the joint line 12 16, between the funnel and spout draw the semi-section E. In this case the spout is a parallel piece of tubing as shown. The first step is to divide the half section of the funnel into equal spaces as shown by the small figures 1 to 11. Note that where the curves are struck with a short radius, as at the ends of the oval between 1 and 3 and 9 and 11, the spaces are smaller than between 3 and 9.

Through the center of the oval draw the lines A B and C D. At right angles to A B and through the small figures 1 to 11 draw lines meeting the line A B, as well as the top line of the funnel in elevation as shown from 1′ to 11′. In a similar manner divide the semi-circle E into equal spaces, in this case four, as shown by

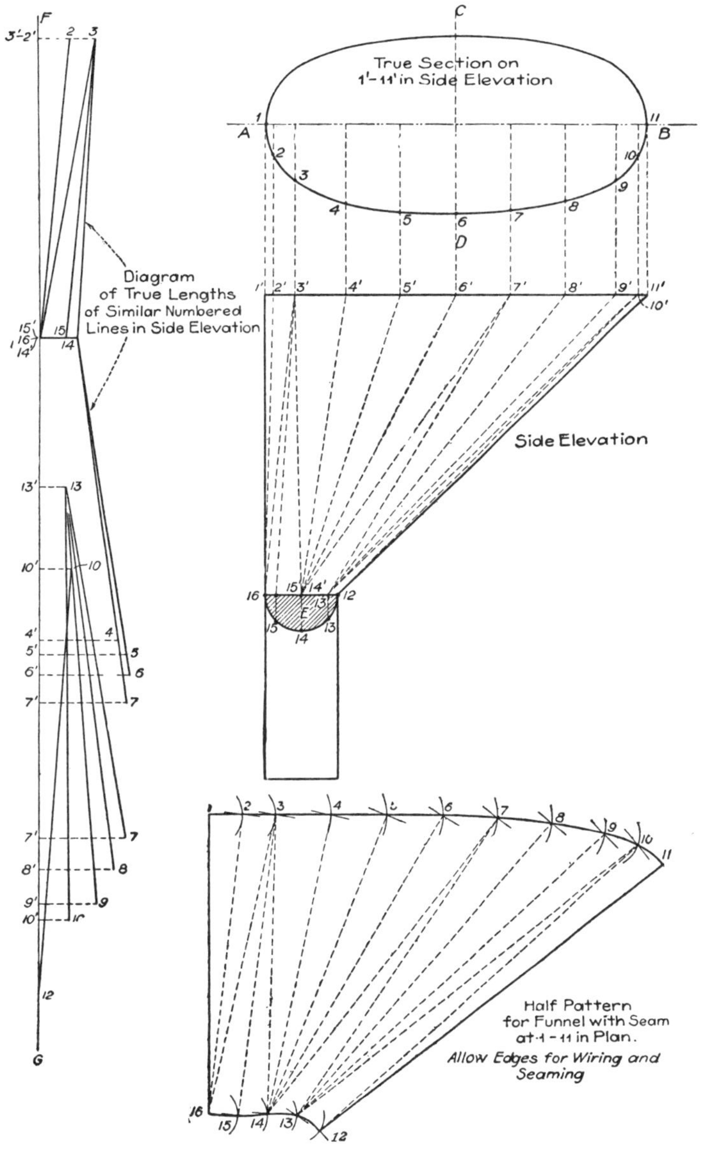

Fig. 156. Elevation, Plan and Pattern for Funnel Perpendicular at One End

the small figures 12 to 16 from which points perpendiculars are drawn, meeting the line 12 16, at 13′ 14′ and 15′. As the semi-oval contains a greater number of spaces than the semi-circle, then connect the various numbered points by lines in a manner as shown in the side elevation. These lines then represent the base lines of sections which will be constructed, with altitudes or heights equal to the various distances measured from the line A B to points 2 to 10 in the half oval and from the line 12 16 to points 13 to 15 in the half circle E, all as shown in the diagram of true lengths. For example, to find the true length of 14′ 6′ in the side elevation, take this distance and set it off on the line F G as shown from 14′ to 6′. From these two points and at right angles to F G, draw the lines 14′ 14 and 6′ 6 equal in length to the distance measured from the line A B to the point 6 in the half oval and the distance of 14′ 14 in the semi-circle E. The length of the line drawn from 6 to 14 in the diagram of the true lengths, will give the true length of the line 14′ 6′ in the side elevation. In this manner all of the true lengths shown, are obtained.

When the half pattern is being laid out the direction of the lines in elevation must be followed so that the lines will run in a similar manner in the pattern. First take the length of 1′ 16 in elevation which shows its true length, and place it as shown by 1 16 in the pattern and with radii equal to 16 2 and 16 3 in the true lengths and 16 in the pattern as center, draw the arcs 2 and 3. Set the dividers equal to the distance 1 2 and 2 3 in the half oval and starting from 1 in the pattern, intersect arc 2, then arc 3. Then using as radii the distances 3 15 and 3 14 in the true lengths and 3 in the pattern as center, describe the arcs 15 and 14, which intersect by the divisions obtained from 16 15 and 15 14 in the semi-circle E. Procced in this manner in completing the pattern, using alternately first the radii in the true lengths, then the divisions in the semi-oval, again the proper radii in the true lengths, then the divisions in the semi-circle, E, until the line 11 12 in the pattern has been obtained, which equals 11′ 12 in the side elevation. Trace a line through points thus obtained in the pattern, and 1 11 12 16 will be the half pattern for the oval funnel with a seam at 1 and 11 in the oval section. Seaming and wiring edges must be allowed to the pattern.

Patterns for a Rectangular Funnel

Fig. 157 shows how the rectangular funnel is laid out. When this funnel is made in large sizes, the four sides are usually seamed together separately, but when the funnel is small, the pattern may be laid out in one with just one corner seamed. Both patterns will be laid out as we proceed. First draw the side elevation of the funnel as shown by A B C D and above this draw the plan view E F G H, which shows the plan view on the line A D. The plan view on B C in elevation is shown by J K L M. Thus the top of the funnel is rectangular while the bottom is square. To the square bottom J K L M the tapering spout is soldered as shown in elevation by Y.

The pattern for the spout Y has not been shown here, as this same pattern was developed in Fig. 155. As D C in the side elevation shows the true length of the ends of the funnel at right angles to F E or G H, it will be necessary to find the true length of the sides, because no elevation of the end is shown. Therefore, take the distance of the projection of the side J *a* in plan and place it from the vertical line erected from B in elevation as from J^1 to a'. A line drawn from a' to B will give the true length of the sides at right angles to F G or E H in plan. From the center *i* in plan draw the horizontal line C° D° and the perpendicular line B° J°. Take the girth from C to D in elevation, and place it as shown by C° D°, through which and at right angles to C° D° draw lines as shown, which intersect by lines drawn parallel to C° D° from the corners E F K and L in plan. Connect the various intersections by lines and then N O P R will be the pattern for the ends. In a similar manner obtain the patterns for the sides. Take the distance of B a' in elevation and place it on the vertical line as B° J°. Through these points B° and J° draw the usual measuring lines, which intersect from the proper corners in plan as shown and S T U V will then show the pattern for the sides.

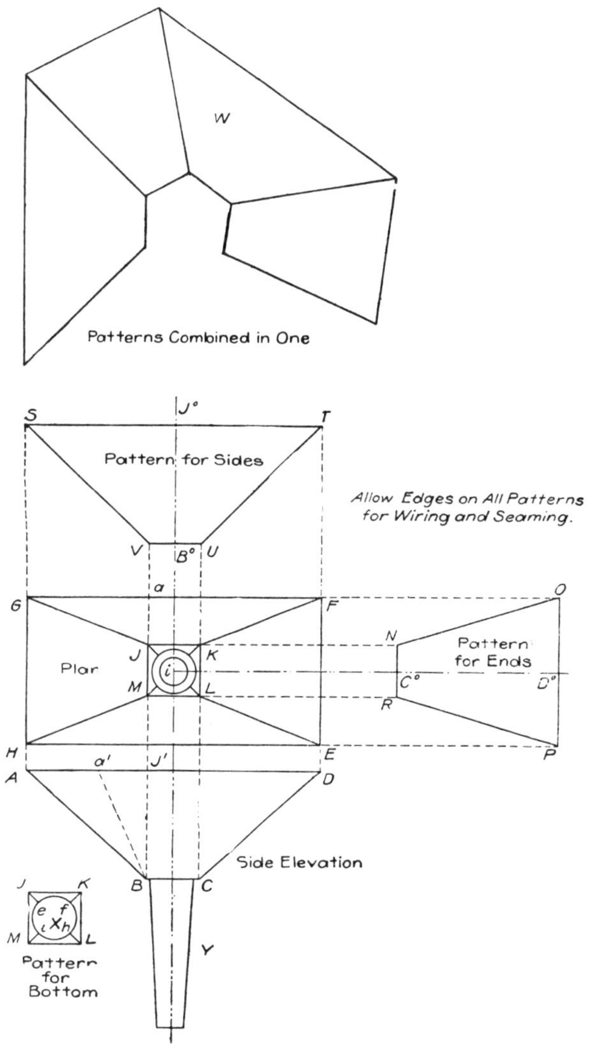

Fig. 157. Elevation, Plan and Pattern for a Rectangular to Square Funnel

If the patterns have been accurately developed, the length of the miter T U in the side must equal the length of the miter N O in the end. If the funnel is small it may be made without waste of material, the side and end patterns being joined in one as shown in diagram W.

In X, J K L M shows a reproduction of similar letters in plan. The circular opening X is cut out, to which the spout Y in elevation is flanged and soldered. Before the bottom X is soldered to the funnel the corners are slightly creased from J to *e*, K to *f*, L to *h* and M to *i* by setting over the hatchet stake with the mallet. Edges should be allowed on all patterns for wiring and seaming.

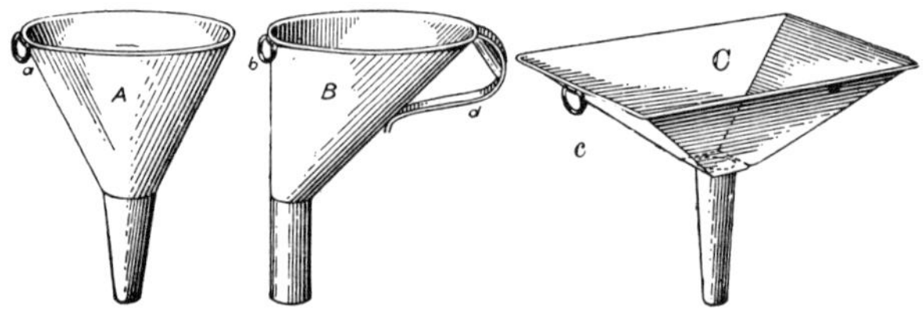

Fig. 158. Finished View of the Three Types of Funnels Previously Described in Figs. 155, 156 and 157

The three types of funnels just described in Figs. 155, 156 and 157 are shown in perspective in Fig. 158 by A, B and C respectively. Wire rings are attached at *a* *b* and *c* or sometimes the handle *d* is attached to the shaped funnel B.

Pattern for Combination Measure and Funnel

In Fig. 159 is shown a finished view of a combination measure and funnel. The method of developing these patterns excepting the handle will be explained in the detail drawing shown in Fig. 160, in which the center line A B is first drawn, and on which the height of the measure C D is set off. On either side of C D set off the semi-diameters of the top and bottom as shown by C 8, C 8^x and D E, D E° respectively. On the top line of the measure draw the side elevation of the funnel, making it of the shape desired, as shown by 1 5 8 9 *i′ i* 13 1. Below the bottom of the measure draw the plan view on the line E E° as shown, the circle being struck from the center A, through which draw the horizontal line *a f* and divide the semi-circle into equal spaces as shown from *a* to *f*. Using C on the upper line of the measure 8 8^x as center, and C 8 as radius, draw the arc, which intersect at 5′ by a line drawn at right angles to C 8 from the established point 5 of the funnel. This gives the true distance from 5 to 5′, which is laid off from 5 to 5° at right angles to 5 1, and from 5° a quarter elliptical shape 5° 1 is drawn at pleasure.

Fig. 159. View of Finished Measure and Funnel

Bisect the joint line 9 13 between the funnel and the spout and obtain 11, which use as a center and describe the semi-circle 9 11° 13, which on account of its small size, is divided into but four spaces, as shown by the small figures 10° 11° and 12°. Through these small figures, at right angles to 9 13, draw lines intersecting the joint line 9 13 at 10, 11, 12. In a similar manner divide the sections

U and T into equal spaces as shown from 1 to 5° and 5′ to 8 respectively From these points, at right angles to 1 5 and 5 8 draw lines intersecting the funnel line

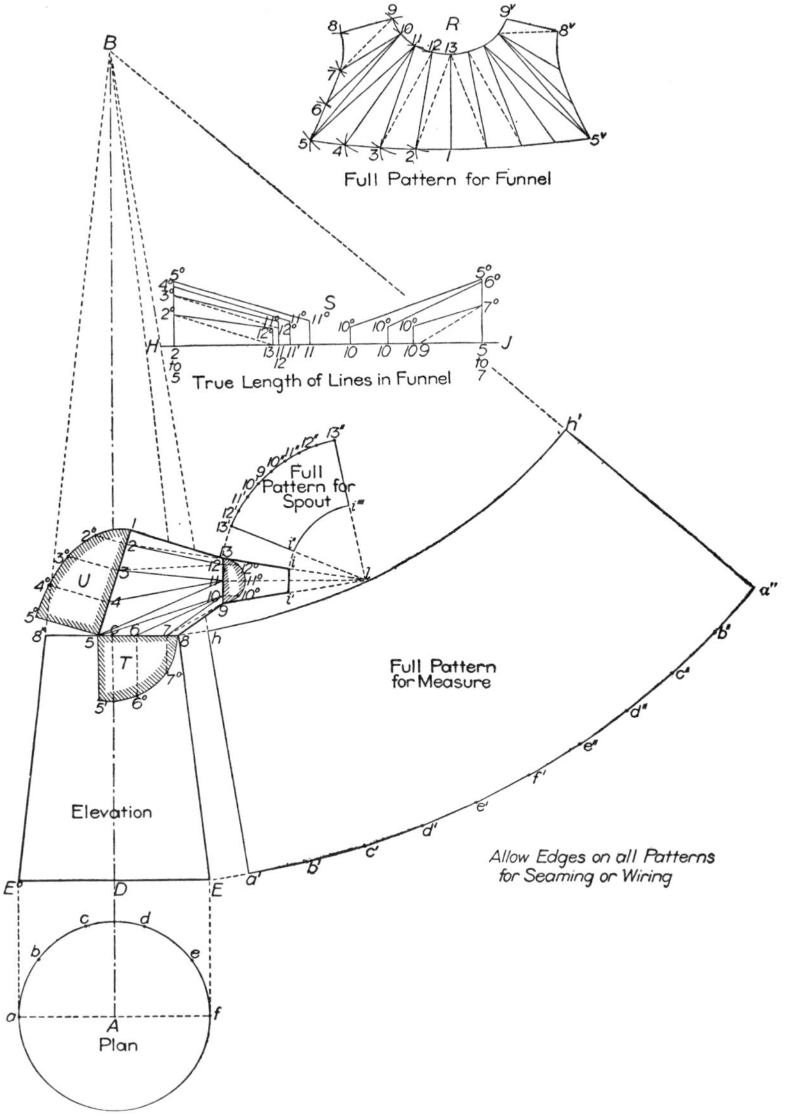

Fig. 160. Details for Laying Out Combined Measure and Funnel

1 5 and 5 8 at 2, 3 and 4, and 6, and 7 respectively. Connect the various points between 1 and 8 to the points between 9 and 13 as shown.

Then will these lines in the funnel elevation, represent the bases of sections which will be constructed in diagram S, the altitudes of which will equal the various heights shown in the profiles T and U, also the semi-circle.

To find the true length of, say, 11 4 in elevation, simply place this distance as shown on H J from 4 to 11′. From 4 and 11′ erect the lines 4 4° and 11′ 11°

equal respectively to the heights 4 4° in the profile U and 11 11° in the semi-circle. A line drawn from the points 4° to 11° in the true lengths S will be the true length of the line 4 11 in elevation. In this manner all of the true lengths are obtained in diagram S.

The pattern for the funnel is laid out as shown in diagram R as follows: Take the length of 1 13 in elevation and place it as shown by 1 13 in R. Now with 1 2° in the profile U as radius, and 1 in R as center, describe the arc 2, which intersect by an arc struck from 13 as center and 13 2° in S as radius. Then using 13 12° in the semi-circle as radius and 13 in R as center, describe the arc 12, which intersect by another arc struck from 2 as center and 2° 12° in S as radius. Proceed in this manner, using alternately, first the divisions in the profile U, then the proper true length; the divisions in the semi-circle in elevation, then again the proper length in S, until the line 11 5 in the pattern R has been obtained. Starting from the point 11 in R, use the divisions in the semi-circle in elevation as radius, then the true lengths in S; the divisions in the profile T, then again the proper length in diagram S, until the line 8 9 in R has been obtained, which is equal to 8 9 in the elevation. Trace a line through points thus obtained, then will 1 5 8 9 13 be the half pattern for the funnel. Trace this half opposite the line 1 13 as shown by 5^v, 8^v, 9^v which completes the full pattern.

The full pattern for the spout is shown by 13′ 13″ *i‴* *i″* and is struck from the center *l*, which is obtained by extending 13 *i* and 9 *i′* until they intersect at *l*. Twice the girth of the semi-circle 9 13 is placed along the outer curve 13′ 13″ as shown.

The pattern for the measure is obtained by extending the sides of the measure until they intersect at B, which use as a center and with B 8 and B E as radii describe the arcs shown. Starting from *a′* on the outer arc, set off twice the division in the plan, as shown from *a′* to *f′* to *a″*. From *a′* and *a″* draw lines to the apex B, cutting the inner arc at *h* and *h′*. Then will *h h′ a″ a′* be the full pattern for the measure.

On all patterns edges must be allowed for seaming and wiring.

This completes the series of articles on Automobile Sheet Metal Patterns; should any question arise which has not been covered in this series, or any new subject which has not been brought out in this book, the writer will be glad to give this additional information if the reader will make his wants known to the editor of The Metal Worker.

BOAT PATTERNS

TWO-PIECE SHEET METAL PRACTICE BOAT

It is not a very difficult job to make a boat shown in Fig. 1, and others can readily follow or modify this plan. To aid them the following description of the methods used for making a small boat of metal, without the use of patterns, and with no special machinery or models is presented for a light, fast boat—a little

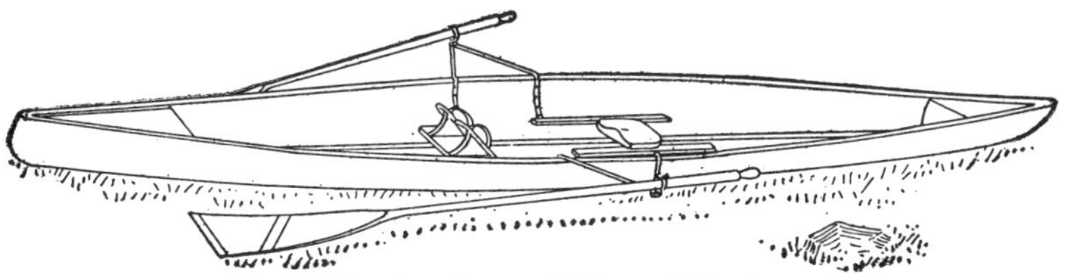

Fig. 1. General View of Boat

cranky, it is true—easily and cheaply made and adapted to either paddle or oars. This boat has proved perfectly safe in use, no defect in construction has appeared.

The present description will have nothing to do with pattern cutting, but will be the method followed in evolving the boat. To begin with, two sheets of No. 27 galvanized iron or steel, 30 inches wide and 96 inches long, should be marked and turned up in the cornice brake, as shown in Fig. 2. This forms the keel and at the same time provides the requisite stiffness for the bottom. Both sheets should now be formed up in the brake or over a large size pipe, to present the appearance of a trough, as shown in Fig. 3. In one end of each trough should be nailed a temporary wood brace about 1 x 4 inches, and of a length corresponding to the beam of the boat, which in this case is 19 inches.

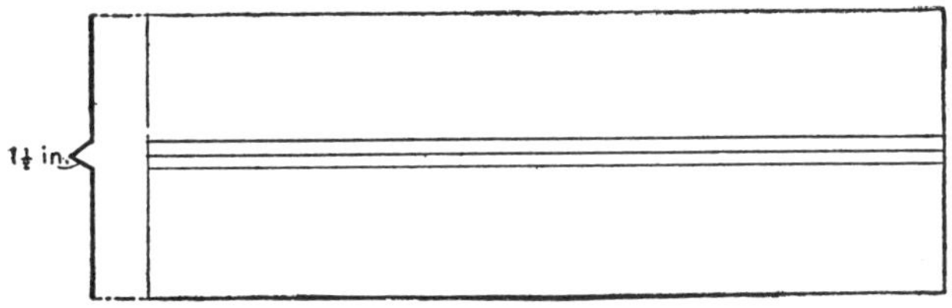

Fig. 2. The First Bend in the Hull Strips

At the end opposite the brace the curve of the trough should be straightened out, and the two sides brought together to form the prows. Each piece will then look like one-half of the boat, but with an exaggerated sheer, and will present the appearance shown by Fig. 4. In this operation care should be taken that the work be done a little at a time on each side, so that the

Fig. 3. Forming the Hull

keel may not be unduly strained, as it is very important that it should remain perfectly straight. At the end of the prow the V-shaped keel should also be drawn together, and both it and the curved sides gradually tapered out to the required size at the middle. The two ends of the prows may now be temporarily fastened together by means of ½ × 3-16 inch stove bolts, and the corner at the bottom cut on a curve, as shown by dotted lines in Fig. 4.

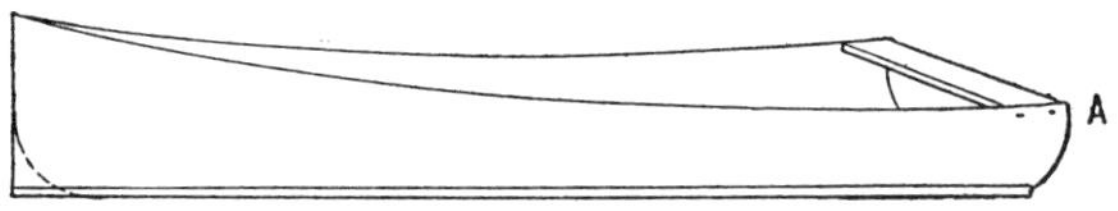

Fig. 4. Forming the Prow

Having determined the amount of sheer desired in the finished boat, the surplus metal should be cut away on a line marked for it. It should be remembered that the amount of sheer may be governed somewhat by the bevel of the sides of the boat at the top of the middle section. If the sides are perpendicular at the point marked A in Fig. 4 there would be no sheer in the finished boat, while the more the sides at that point are inclined outward the more sheer will be produced. In the present boat the sheer is only 2 inches—that is, the depth at the prow is but 2 inches more than the depth at the middle section. The sides of the boat may, therefore, be practically perpendicular.

Fig. 5. The Prow Piece

The metal at the prow should now have all the buckles hammered out, which may be done by using a smooth faced hammer on the inside and holding the square head on the outside, thereby stretching the metal. The prow ends may then be riveted together permanently. A piece of sheet iron should be cut the shape of the prow, as shown in Fig. 5, about 1 ¼ inches wide, and so placed that the outer edge should project about ¼ inch beyond the metal of the sides at the prow. Removing the temporary stove bolts, this piece should be riveted, together with the two thicknesses of the metal at the prow, with rivets ¾ inch apart, and the projecting edge of the separate piece clinched over the raw edges of the prow. The seam should then be thoroughly sweated with solder. The air chambers should extend about 28 inches back from the points of the prows, and should be air tight They are to prevent sinking should the boat fill with water. A piece cut the required size with 1-inch flanges, and extended as shown at C in Fig. 6 should be bent down at the dotted line A B. This forms the top and one end of the air chamber, and should be riveted to the sides of the boat, so that the top of the air chamber will be 1½ inches below the top of the boat. The extension C, Fig. 6,

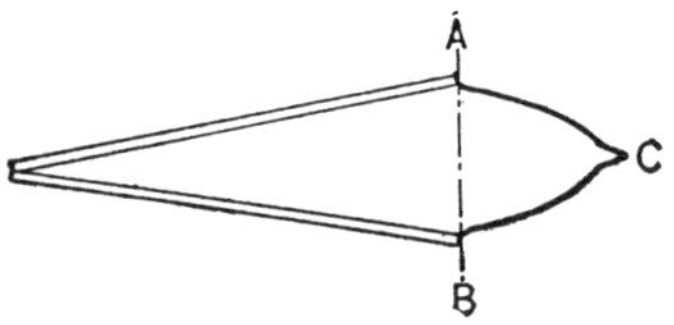

Fig. 6. The Top of the Air Chamber

should be long enough so that it fits snugly in place, inclined at an angle, and should be carefully soldered to the sides of the boat, instead of being riveted. When this is done, the two parts of the boat (which should be alike) may be fitted together at the middle.

Two boards with straight edges 1 inch apart should be nailed to a level floor. The keel may be set down in the groove thus formed and a level used on the wood braces. A few blocks nailed to the floor and marked at the hight of the sides will enable the boat to be removed from these primitive stocks and replaced exactly as before, without frequent recourse to the level and square. The two ends should be accurately fitted together with a lap of 1 inch and lightly tacked together with solder. The boat should then be turned over and a chalk line stretched along the

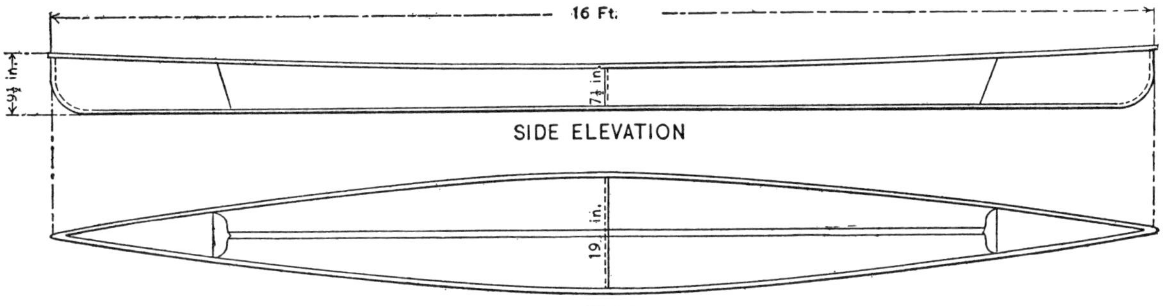

Fig, 7. Plan and Side Elevation of Two Parts Connected

keel. Any inaccuracies should be corrected, and the middle connecting seam securely riveted with rivets ¾ inch apart and soaked with solder inside and outside. The boat will then present the appearance shown in Fig. 7, which gives both the plan and side elevation with dimensions.

The topsides, or gunwale, is composed of two strips of white pine with the sheet iron between, and screwed together, with screws about 4 inches apart. The inner strip which is ⅜ × 1½ inches, receives the heads of the screws. The outer strip is ¾ × 1½ inches and is made thicker in order to more securely hold the tails of the screws, as shown at A in Fig. 8. The wood at the extreme end of the prows should be rounded at the points and bound with sheet copper fastened with small brass nails. To this copper may be soldered a small tube, in which may be inserted the staff of a small flag.

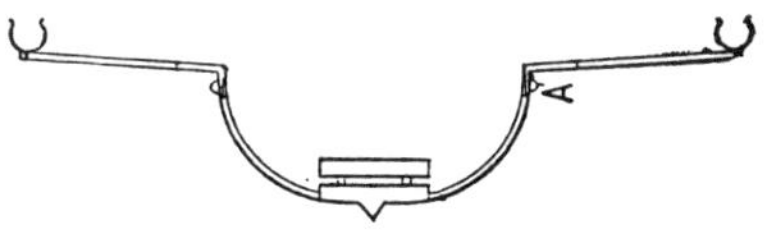

Fig. 8. Cross Section, Showing Gunwale Strip, Bottom Board and Oarlock Outriggers

The boat, or canoe, at this point weighs 35 pounds, and with a double bladed paddle forms a very satisfactory fast canoe for use in comparitively quiet waters. It may be painted to suit the individual taste. To transform it into a practice boat the seat and oars may be added. These are entirely separate from the shell. A white pine bottom board 1 foot wide, 4 feet long and 1¼ inches thick is shaved to fit the bottom of the shell near the middle section.

A pair of irons should be prepared by the blacksmith to fit loosely inside the shell at that point, and extend out over the sides 12 inches, joining together at the ends to receive the swivel oar locks, as shown in Figs. 8 and 9. These irons are ⅝ inch round where they overhang the water and are flattened on the inside of the boat. The under side of the bottom board is gouged out, so that the irons set in the wood, and the irons are securely bolted to the board, taking care that the bolts are countersunk, so that they do not scratch the shell of the boat. A foot rest with braces and toe straps should be securely screwed to one end of the board, and guiding strips for the seat rollers fastened in about the position shown in Fig. 8. Inside the guide pieces, where the wheels are to run, two strips of hard wood are inlaid to take the wear.

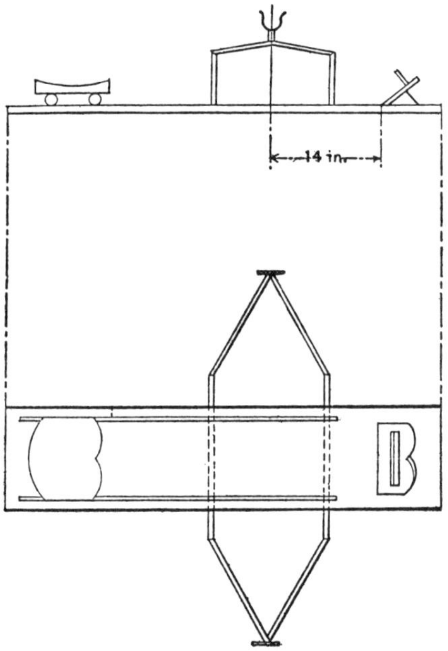

Fig, 9. Plan and Side Elevation of Seat and Outriggers

The roller seat should be gouged out of a piece of white pine about 7 inches square, and provided with rollers, which may be procured at any hardware store. It should be borne in mind that the lower the seat and rollers are placed the steadier the boat will be. With such a removable seat the position in the boat may be varied to suit, and as both ends of the boat are alike, the bottom board and seat may be reversed without necessitating the turning of the boat. Feathering spoon oars should preferably be used, although it takes a little practice to become accustomed to the proper turning of the wrist.

The cost of the material for the shell alone, including paint, should not be more than $2.50. The time and trouble will be more than repaid by the pleasure of one season's use. The keel, as shown, would be insufficient for an ordinary boat, but for such a boat it is ample, because the boat, when loaded, acts as a keel for itself. It may be added that on one occasion a man weighing 150 pounds shipped so much water into the boat that it sank to a point where the water was half way between hip and shoulders, and the boat was then rowed ashore in the usual way, except that the boat itself was entirely under water.

CONSTRUCTING SHEET METAL BOATS ON SKELETON FRAMES

In shops doing this kind of work sheet metal boats are made in various sizes, for which there are kept on hand skeleton frames of wood, giving the length and beam of the boat and the various profiles throughout its entire length. These frames are used over and over again if a boat is required of that particular size, or,

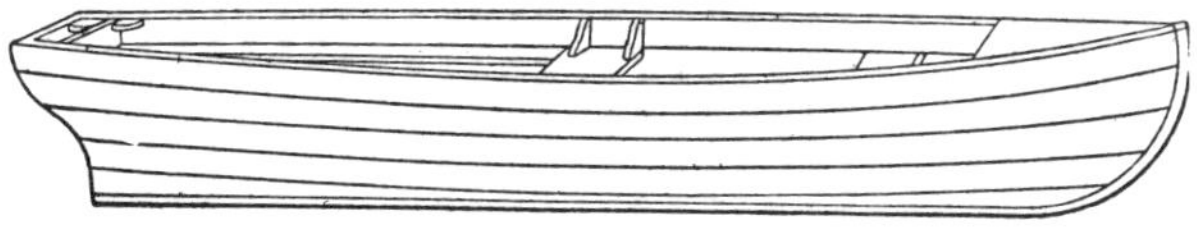

Fig. 1. General View of Boat

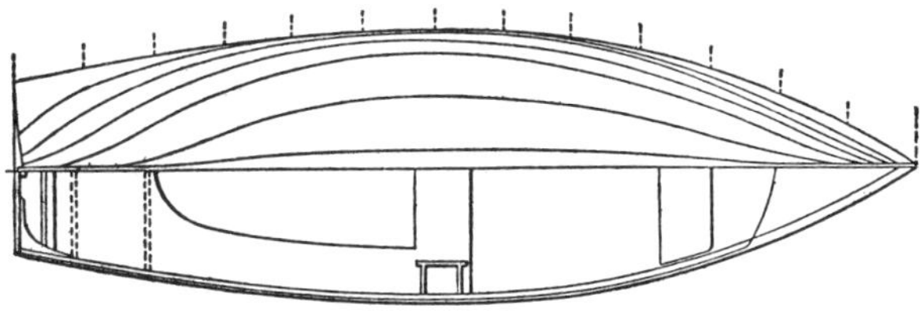

Fig. 2. Plan View

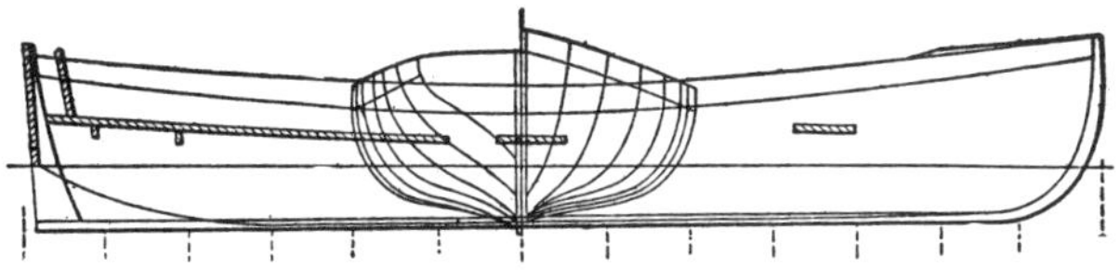

Fig. 3. Longitudinal Section and End View

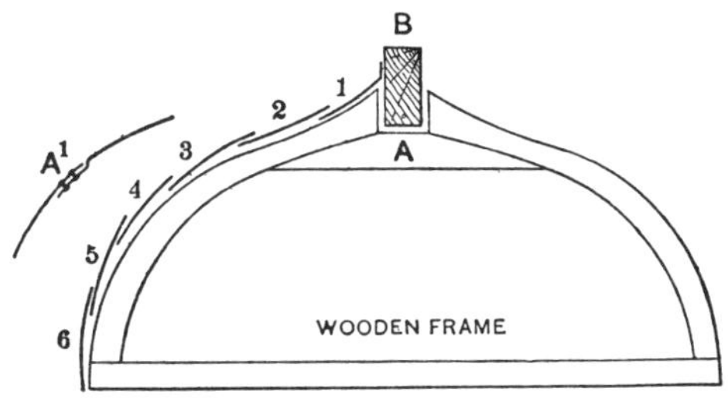

Fig. 4. Section of Skeleton Frame Made of Wood

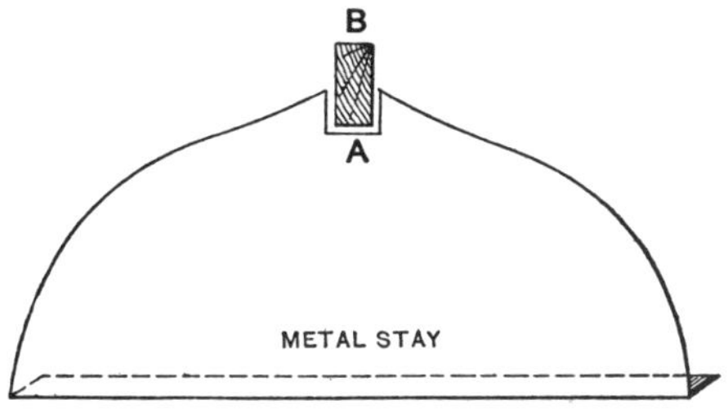

Fig. 5. Section of Metal Frame

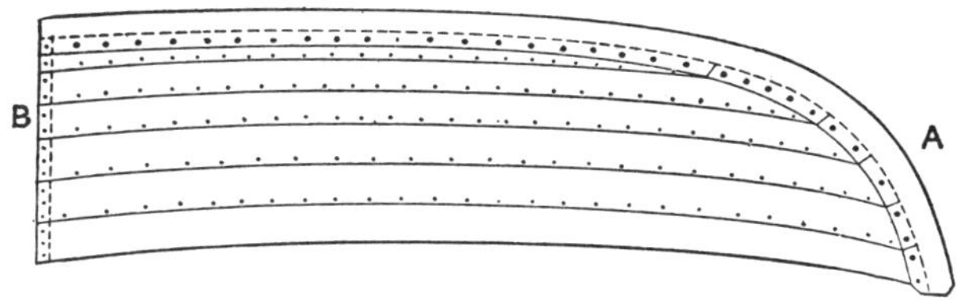

Fig. 6. Showing Gore Piece Flanged Out Against the Keel

PATTERNS FOR SHEET METAL BOATS

if a special size is needed, a new frame is made, and used as often as required. In this way the various shapes accumulate and are kept in stock.

If, however, it is desired to make but one boat for pleasure use, a skeleton frame work of metal stays is built up full size. First draw the plan view of the boat, as shown in Fig. 2; then draw and cut out of sheet metal the various shapes or sections, full size, as shown in the end views in Fig. 3. These stays should be soldered in their proper positions on the plan of the boat as in Fig. 2, and braces placed between the stays to hold them in a vertical position. Now, whether the skeleton frame is made of wood, as shown by the section in Fig. 4, or of metal, as shown in Fig. 5, a notch must be cut at the top, as at A and A in Figs. 4 and 5, to receive the keel of the boat, B.

This keel is made of hardwood and covered with No. 22 galvanized iron, nailed with galvanized iron or brass nails, and the seams and nail heads thoroughly soaked with half and half solder, using 10-pound soldering coppers for same. The keel is then scraped well to obtain a smooth surface, when it is laid into the notches, as shown in Figs. 4 and 5. The keel being in position, the skeleton of the boat is placed upon a number of wooden horses or brackets, and the patterns for the gores are obtained. Space off on the stays of the frame the number of gores required; in this case six, as shown in Figs. 1 and 4. In other words, as each profile or stay varies in girth, each one must be divided into six equal spaces. From these spaces run a chalk line or soft, thin copper wire from stem to stern, being careful that the line touches each division on the stays. Then take some manila paper or very thin tin foil and place it over the lines drawn, when an accurate pattern is the result. These six gores, each being a different shape, are cut from No. 22 galvanized iron, allowing laps for riveting.

The metal work is now constructed over the skeleton frame, starting on the bottom at the keel, as shown by the gore piece 1 in Fig. 4. This gore piece is flanged out 1 inch against the keel, as shown in Fig. 6 from A to B, and nailed with galvanized iron or brass nails 1 inch long, and then thoroughly soaked with solder, using heavy roofing coppers. The first gore piece will require some stretching with the hammer to make it conform to shape desired, while the following gores 2 to 6 in Fig. 4, are formed up on a small roller similar to that shown in Fig. 7, C showing the section. When a large number of various size boats are made, various size male and female rollers are kept in stock, and placed in the standards A and B; and by means of the handle shown, the gores are fed through the rolls until the desired shape and radius are obtained. If a smaller or larger boat is being built, the rolls are taken out of the standards and the required rolls replaced. Where, however, but one boat is required the gores would have to be hammered up by hand, by using the required size raising hammer and a block of wood or part of the trunk of a tree about 12 inches diameter and 3 feet high, or on a block of lead. Assuming that this has been done, gores Nos. 2 and 3 in Fig. 4 are then placed in position, as shown in Fig. 8. By using a hand clamp the gore is tightly fastened against gore No. 2. But before putting gore 3 over gore 2 take a brush with thick metallic paint and paint the entire seam of 1½ inches on gore 2 or any other number. On top of the painted seam put a strip of asbestos paper about as thick as ordinary blotting paper, on top of which place gore 3. When riveting use 2-pound rivets, tinned, and be sure that the holes punched will be no smaller or larger than to just allow the rivet to fit snugly. Then rivet in a manner as shown in Fig. 9, always placing the top and bottom rivets between each other so close as to insure a tight joint. In this way all of the gores from 2 to 6 are fastened.

Where the gores are placed over sheet metal sections as in Fig. 5 some provision must be made for fastening the seams, as they cannot be fastened with the hand clamp. The seams can either be tacked temporarily with solder or wooden

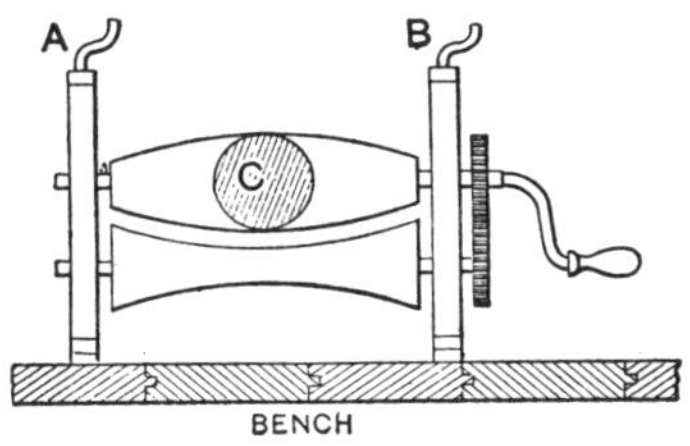

Fig. 7. Side View of Small Rollers

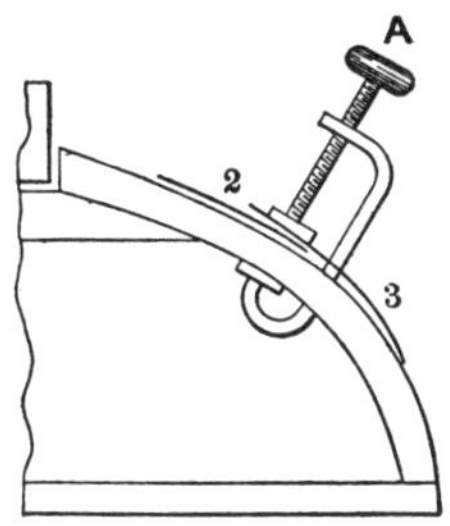

Fig. 8. Showing How the Gores Are Fastened in Position

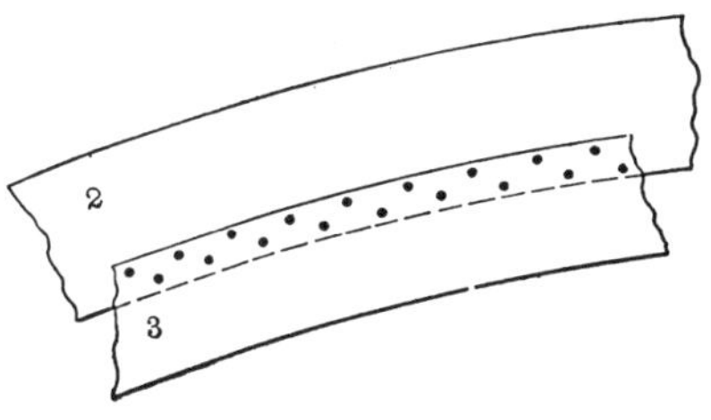

Fig. 9. Showing Manner of Riveting

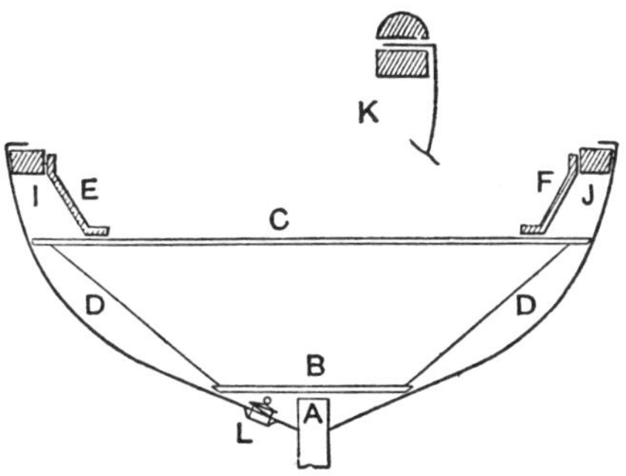

Fig. 10. Section Showing Manner of Fastening the Hard Wood Rail in Place

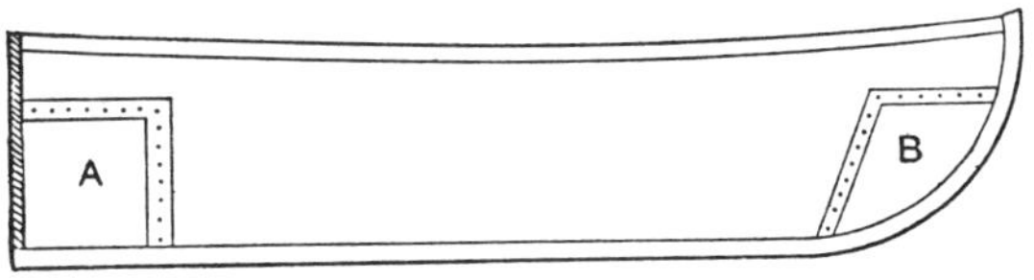

Fig. 11. Showing Position of Two Air Chambers

PATTERNS FOR SHEET METAL BOATS

lookouts put in place. After the gores are riveted water tight, then where each gore meets the stem and stern, as in Fig. 6, they are flanged so as to be nailed with galvanized or brass nails, and thoroughly soldered with heavy coppers, as at A and B. The stern of the boat at B is made of hard wood entirely covered on both sides with galvanized iron nailed and soldered in the same manner as described in connection with the keel. This, of course, is done before the gores are put in position.

The boat being finished thus far, the seams are set down so as to have a smooth surface on the outside, as shown at A^1 in Fig. 4, then the hardwood rail is placed at the top, as shown at I and J in Fig. 10, flanging the metal over the top as shown, over which a half bead is placed to make a smooth finish, allowing spaces for the row locks, as shown in K. In the bottom of the boat a tapped ferrule L, made of brass, is soldered, into which a threaded socket fits; then when the boat is in use and some water should accumulate in the bottom, it can be easily drained when the boat is not in use by unscrewing the socket, as is shown at L in Fig. 10. When the keel projects on the inner side of the boat, as at A, this forms the rest for the platform or foot rest B. C shows the seat resting upon the brackets D and D. Band iron braces E and F, galvanized and made of 3-16×1 inch band iron, are fastened to the rail and seat by means of wood screws as shown.

If desired, two air chambers can be put in the boat, one at the stem, the other at the stern as shown by A and B in Fig. 11, which will at the same time form a seat. These air chambers should be made of No. 20 galvanized iron, well riveted and soldered. A rudder can be placed at the stern by placing two eyes in the stern, onto which the rudder is hung by means of the two pins in same. In this connection it may be proper to say that if our correspondent desires the boat for his own use, and he can obtain a wooden boat which has passed its usefulness, he can strip off the outer covering, thus leaving the frame, which will be an excellent model on which to construct the metal boat.

BUILDING METAL BOATS BY MEANS OF MODELS

The first thing to do in the building of boats is to make a small model on a scale of 1 inch to the foot, of plaster of paris. To prepare this model, take a box of about the size of the proposed model and fill it with plaster of paris, mixed quite thin with water and stirred well. After it has settled take the box apart, dry the plaster thoroughly and carve out the model.

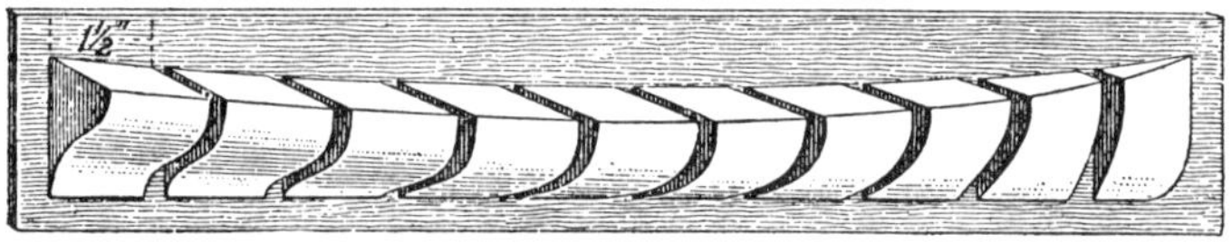

Fig. 1. Model Cut Into Sections

The next step is to cut the model into sections. This is done with a saw by cutting perpendicularly through the center from end to end and on a line with the keel. Take each half of the model and cut it across in sections 1½ inches long. Fig. 1 shows a sketch of a model cut up in this way. The next step is to take each slice of the model and section it, as shown in Fig. 2 in reduced size, laying out the lines equal distances apart and at right angles to one another, making the squares about ¼ inch on a side. Now lay out the section full size, as shown reduced in Fig. 3, to correspond with the model. If the largest cross section in the model is 4 inches, the plan should be laid out 4 feet, since the scale, as stated above, is 1 inch to the foot.

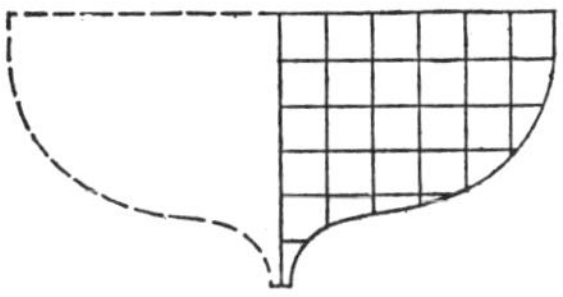

Fig. 2. Cross Sectioned Model, Reduced

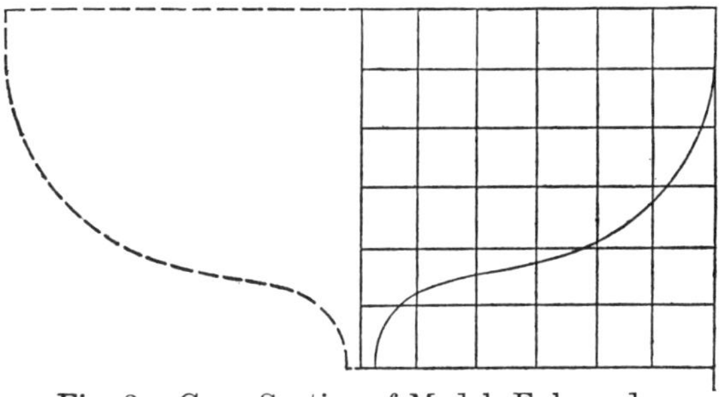

Fig. 3. Cross Section of Model, Enlarged

In the diagram, Fig. 3, the same number of spaces is laid off as on the model. Next trace the section line or profile through the spaces, which can be done by transferring the points where the profile of the model cuts the sections in Fig. 2 to corresponding points in Fig. 3. Repeat this operation with each of the sections shown in Fig. 1. After having a full sized profile of each section, cut them out of 1-inch boards. Set the boards or profiles 18 inches apart, or whatever distance is right to correspond with the scale of the model, and fasten these profiles firmly together. If the boat is a keel or round bottom boat, fit and fasten the keel at the proper place to the wooden stays. The keel can be made of iron or hardwood. Now draw horizontal lines on the edges of the stays or frames and these lines will be a gauge to which to cut the strips for the body of the boat.

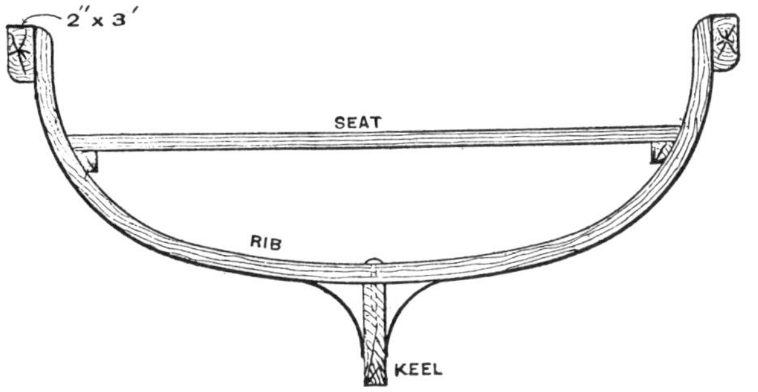

Fig. 4. Section Through Finished Boat

The easiest way to cut the strips is to lay the iron on the stays and mark out the strips to correspond with the lines. The width of the strips depends on the shape of the boat, and the builder must use his judgment in deciding what width to cut them to. Nail the strips to the stays and rivet them together between the stays. If the ends of the boat are pointed and it is impossible to get between the sides to rivet properly, take out the keel and cut the stays or frame in two from end to end and then rivet the ends. Then fasten the stays together again and nail or rivet the keel to the iron. Take a strip of hardwood about 2×3 inches, or angle iron and fasten it around the top edge, and then put in the ribs, which can be of wood or light angle or tee iron, as preferred. The next operation is to take out the stays and solder the boat on the outside. A cross section through the boat is shown in Fig. 4. Make each end of the boat air tight and put water tight boxes under the seats, which will make a life boat of it, as the air compartments will prevent it sinking in case of filling with water. The first two coats of paint should be of red lead.

SHEET METAL CANOE

The following is a description of a sheet metal canoe: To begin with, it is made of six sections of No. 27 galvanized iron, 26 in. wide, straight and level on the bottom: It was made bottom side up, and was molded after the following

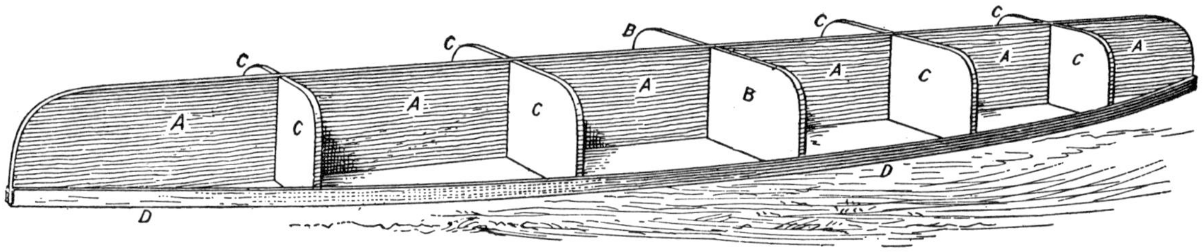

Fig. 5. Mold or Frame for Forming Canoe

fashion: A board 14 in. wide was taken of the desired length, in this case about 12 ft. long. This was fastened to the floor, and the shape resembled the sketch presented in Fig 5, in which A A., etc., shows the length and depth of board. The width of this board governs the depth of the boat, but the length can be varied to suit the material and desire of the maker. At the center of this long piece is connected two short pieces, B B, each half the width of the boat, and of the shape indicated, with one corner rounded to whatever circle best suits the ideas of the builder. For the outside edge of the bottom of the boat a 3-in. circle was used. After these pieces were nailed to the centerpiece, thin, narrow strips of elm, D D, were taken about $\frac{3}{4}$ in. thick and $1\frac{1}{2}$ in. wide, long enough to spring around the short cross pieces B B, and were attached to the long centerpiece A A at each end. These pieces were bent so as to give the desired shape to the gunwale of the boat. The distance between the four other short cross pieces C C, on which and over which the sheet iron is formed, is governed by the width of the sheet iron used.

In this boat, 26-inch iron was used. When the pieces C C are securely fastened to A A and D D the mold for the canoe is ready for forming the sheet iron sections and riveting them together. If 26-inch iron is used it will allow for sections 25 in. long, after the laps are made. The shape of the finished canoe will be governed by the shape of the frame or mold. The bow and stern sections will have to be made in two pieces, with slits, as indicated at E, and made as designated in Fig. 6, to be covered by laps or patches to be riveted on and soldered. These slits aid in forming the bow and stern. The ends are also slitted or notched to aid in forming a strong, water tight joint at the bow and stern. When the mold is securely fastened together there is some advantage in having it so it can be lifted up and set on boxes, so as to avoid stooping to work on the floor It would be necessary to set the mold up if a canoe was being made with some shear.

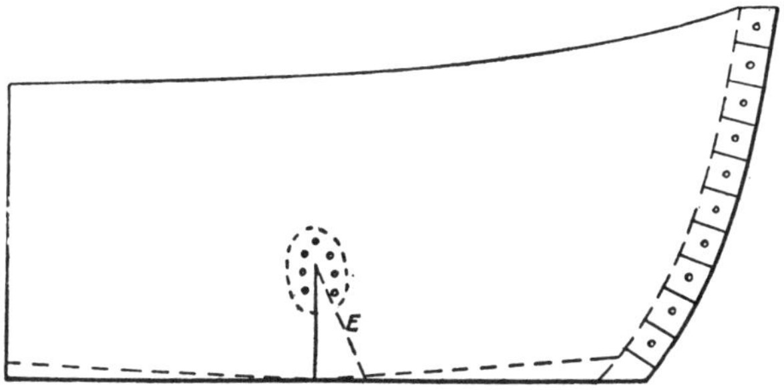

Fig. 6. Pattern of Bow and Stern Pieces

The work can best be started at one end, preferably the stern, to which the skeg is attached in the center and in line with the bow. The stern and stem pieces shown in Fig. 6 should be notched about 1 in. apart. The other side should be notched in the same way, but the two parts should be connected, so as to break the joints and alternate; the notched parts turned down so as to come even with the extending piece on the other sections, so that they can lap over on the section proper and be riveted in place, and finally soldered heavily by soaking. In order to give a little fullness to the canoe and a greater carrying capacity these end pieces should have a slit cut about midway of the section, so that one edge of the cut can be lapped over the other, and the part drawn until the straight line overlaps the oblique line shown at E, in Fig. 6, where it should be riveted in place. This will make a hole at the top of the gore, over which a patch must be riveted and soldered, as shown by the dotted lines. This will naturally throw the bottom line of the bow up a little, or out of a straight line with the bottom, and will necessitate its being trimmed so as to be on a straight line with the keel. A three-cornered bottom piece will be needed to connect with the side pieces, as the practical man will learn and supply.

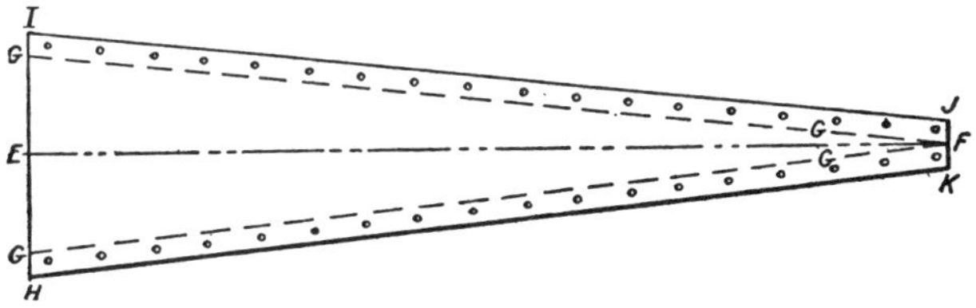

Fig. 7. Pattern of Skeg

1-lb. tinned rivets are used, placed ¾ in. apart, putting the heads on the outside always, and soldering all over the rivets well on both sides; also the seam. When the stern section has been snugly fitted temporarily to the first mold board C C, then the next section of sheet iron can be fitted, which is comparatively easy. When it is snugly fitted the next section is made, and finally the bow section completed. Care must be taken to fit each section snugly and true, and if this has been done all should be marked and the sections riveted together and soldered. When this is done the shell of the canoe is ready for reinforcement and the necessary fittings.

Good strips of elm or any other strong wood that will bend readily should be used for the gunwale to stiffen the canoe, and they should be made 2 in. wide and from ¾ to 1 in. thick. Fit these on the inside and nail into position with clout nails and clinch them on the inside. At the bow and stern fit in a piece or two of wood of proper shape to which fasten chest handles for carrying the canoe, hanging it up when not in use, and for attaching a line for mooring it. One can carry this canoe easily Indian fashion, bottom side up with the gunwales resting on his shoulders. After these end pieces have been put in, a brace should be put across the canoe about one-third the distance from each end, using ½ in. iron pipe with a blank flange on each end drilled with screw holes for fastening to the gunwales. These braces, with the gunwales, keep the canoe from collapsing, the sheet iron being strong enough otherwise to hold the shape of the canoe without the need of ribs. This canoe is provided with outriggers for the purpose of rowing. For outriggers, 12-in. T hinges are used fastening the T part to the gunwale with stove bolts. The long or strap part of the hinge will then fold into the canoe out of the way when not in use. On the outer end of the strap are bolted U shaped pieces of iron strong enough for row-locks.

On the outside of the canoe at the stern, and in line with the keel and bow, a skeg should be fastened. The shape will be something like Fig. 7 before it is formed into a sharp V on the line E F, and the edges turned on the lines G G after holes are punched for riveting to the bottom of the stern piece. The skeg will be stronger if the V is not too close at the point of connection with the canoe bottom. The holes in the canoe bottom for the rivets are easiest made with a twist drill of the proper size, using the skeg for a template to insure the holes being in exactly the right place and to avoid forcing the skeg out of line. The rivets should be put in the holes H I J K first and the others afterward. It is easier to solder the skeg to the bottom in proper alignment and drill the holes afterward. The rivet heads should be on the outside of the canoe, thoroughly and smoothly soldered. The skeg is about 30 in. long and $2\frac{1}{2}$ in. deep at the wide end. It should be made smooth at the point where it is attached to the canoe so as to avoid catching grass or anything to cause trouble. If attached on the proper line and made correctly it will greatly aid in keeping the canoe on a straight course when in motion.

Two cypress boards are fastened together and fitted to the bottom to support the seat box. There are no nails in these boards where they can scratch the metal or possibly wear a hole. The seat box is 26 in. long, 12 in. wide and 8 in. deep. It is covered with tin and soldered air-tight. This has two wire rings soldered on each side and is fastened by means of cord to staples in the wooden bottom. Although none were used it would perhaps be better to place air chambers in the ends.

BUILDING SHEET METAL BOATS FROM GORES OR STRIPS

With the constantly increasing scarcity of wood and the consequent advance in the price of lumber, there has been a growing tendency for a number of years to substitute other materials where formerly no one ever considered the use of anything but wood. The principal of these displacing materials are the iron and steel products, and accordingly sheet metal is used at the present time for a greater variety of purposes than ever before. One line in which its economy and practicability has been demonstrated for a number of years, and one in which it might be used to a greater extent, is in the building of small boats. As far back as 1850 a sheet metal boat known as the Francis Patent Metallic Life Car was used in life saving work off Squam Beach, N. J.

The accompanying illustrations will give an idea of the plans of constructing two sheet metal boats followed by two men after regular shop hours, one doing the metal work and his friend attending to the wood work. Each thus became the possessor of a boat at but a slight cost in cash paid out over the actual cost of the materials used.

As soon as the patterns were completed from sketches, work on the boats proper was started. They are 15 ft. long, 42 in. broad, 15 in. high in the center and 22 in. at the bow and stern. Each boat as shown has a sharp bow and the stern is pointed below the water line, flaring out as the top is approached until it is practically square. The keel is made of white oak $1\frac{1}{4}$ in. thick and $1\frac{1}{2}$ in. deep, and dressed. A triangular piece of wood $1\frac{1}{4}$ in. thick and about 3 ft. long was nailed on top of the keel at the stern to form the skeg, and also a support for the board which was used to form and reinforce the stern. The stem post was then fitted on to strengthen the bow. It is made of hardwood and forms an extension to the keel. Both the bow post and the stern board are covered with metal of the same weight as the shell of the boat.

The shell is made from five pairs of gores or strips of metal cut from 8 ft. stock, with joints broken amidships. The first two gores, marked 5 in the drawings, one on each side of the keel, were flanged, so that they formed up against the side of the keel and lapped across the lower edge as shown in the cut. At the stern these strips were drawn up from the keel, leaving it bare and forming a covering for a portion of the skeg. Two sheets were then cut to extend from these strips down

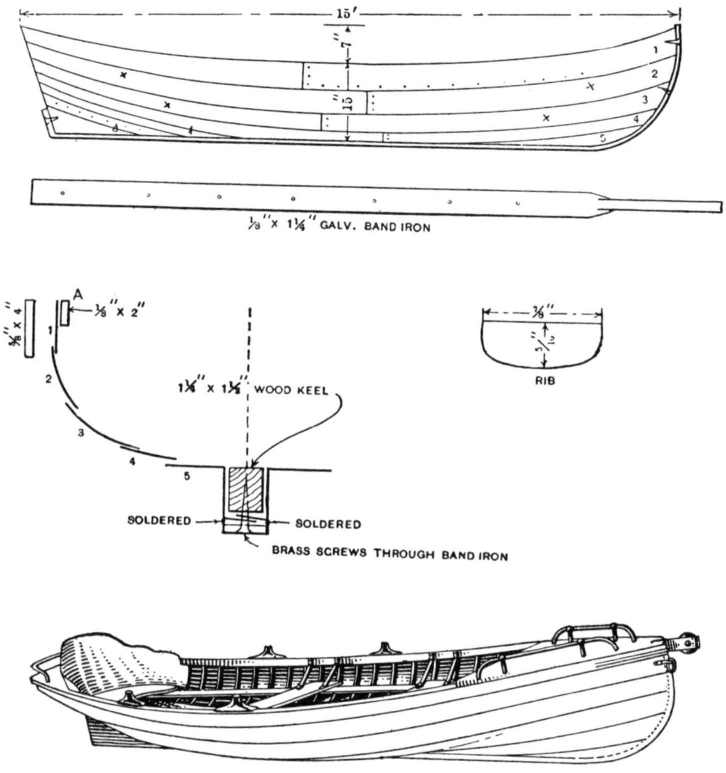

Fig. 1. A Sheet Metal Boat and its Various Parts

and lap under the keel. These sheets were then riveted through the skeg. A band of iron $\frac{1}{8} \times 1\frac{1}{4}$ in. was laid on the under edge of the keel, lapping on to the rear of the skeg and fastened to it by brass screws $1\frac{1}{4}$ in. long, spaced a few inches apart and countersunk. At the bow this band is rounded and curves up to the bow post as a support. The seams between the metal strips and the iron band were soldered to make a water tight joint.

The keel was marked off into 5-in. spaces for the location of the ribs. The next pair of gores, 4, were cut and placed in position with practically no forming. Each strip was fastened to the lower one by tinned rivets, there being three in the space between each pair of ribs. The next two pairs of strips had to be raised on a block with a raising hammer, on the first pair, the raising being commenced about 3 ft. from either end of the boat and on the next about 2 ft. from either end. The amount of the raising was determined by the eye and by trying, and did not take over two hours' labor for each boat. The fifth and last sections, No. 1, were then cut, and as in the previous cases riveted to the lower gores. As the different sections were put in place they were lapped around the bow post and riveted through and also through the stern board. Special care was taken to have the rivets neither too large nor too small for the holes, and they were staggered with the row on the gore below.

Ribs of elm 7/8 in. wide and 5-16 in. thick were placed in position on 5-in. centers and were fastened to each gore by a copper rivet and washer and nailed to the keel. These served to give ample stiffness to the boat. A strip of wood 5/8 × 4 in. was placed around the boat flush with the top on the outside, and a metal strip 1/8 × 2 in. on the inside to form the gunwale. A 1/2 in. bead was then placed around the upper edge, leaving spaces for the oar locks. A metal air chamber was constructed in each end, and afterward all seams soldered on the outside of the boat. A removable board was placed in the bottom to protect it and held in place by small clips on two longitudinal stays. Another longitudinal stay was next placed in position on each side to form a support for the seats, which were made from 1-in. lumber and braced by light iron strips. The ends of the boat were decked in and the stern seat put in place. By way of giving the boat a finished appearance, brass rails were put on the decks, and for convenience in hitching the boat a ring was attached to the bow. The boats were now ready to be painted. The first two coats consisted of red lead and the finishing coats to suit the taste of the owner. In constructing such boats, patterns may be dispensed with if the sheet metal worker can obtain some discarded wooden boat of the desired shape, or if such is not available and he does not desire to make his own patterns from scale drawings they can be purchased from companies in that line of business.

PATTERNS FOR A SHEET METAL BOAT DEVELOPED BY TRIANGULATION

To develop the patterns for the various gores in a sheet metal boat, when the seams run longitudinally proceed as follows: In laying out the patterns in Figs. 1 and 2, five gores have been used in the construction of the boat, so as to avoid a confusion of lines in so small a drawing. In practice, however, any number of gores can be used, applying the principles which will follow.

Let A B C D, Fig. 1, represent the side elevation of the boat. Establish where the extreme width of the boat shall be, in this case on the vertical line 9 57, and parallel to this line draw the center line E F at the left and E° F° at the right. Set off from these two center lines the half width of the keel, as shown at the left and right by v 57° and v° 57° respectively. In line with 9 57 in the side elevation draw the normal or given profile 9° a 57°, or as shown by P at the left and P° at the right. This given profile forms the basis, from which the various proportional vertical sections will be obtained. From a the widest part of the profile P, draw a horizontal line intersecting the line 9 57 in side elevation at a'. From A through a' to D draw the dotted curved line, which will represent the extreme outer edge of the boat in elevation.

The drawing of the plan is now in order. Draw any horizontal line as G H, on either side of which place a duplicate of the half section P, as shown in plan by 9 a 57 on both sides. Project the line 9 57 in elevation and intersect it by horizontal lines drawn from a and a in the true section, thus obtaining a and a'' in plan. Extend the keel lines from 57 in the true sections and intersect them by lines projected from A, b and D in elevation as shown by b b', A A^1 and D D^1 in plan respectively. From 9 and 9 in the true section, draw horizontal lines intersecting the line a a'' at 9 and $9''$ respectively. Draw a graceful curve from A to 9 to D in plan, which will be the plan view of the boat when looking down on the line A D in elevation.

In a similar manner draw a graceful curve b a D in plan, which represents a horizontal view on the line b a' D in elevation. Continue a graceful curve from D a in plan to A as shown by the dotted curved line a A. Then A a D shows a horizontal view of the extreme outline of the boat when viewed from the top, on the line A a' D in elevation.

These lines A a' D in elevation and A a D in plan are used for obtaining the extreme widths of the proportional sections. If desired, the lower half plan shown by b' a'' D^1 and D^1 $9''$ A^1 can be duplicated from the upper half, although this is not necessary in the developments of the patterns. In this case we will assume that the boat is to be made in five gores, as shown by the profile P at the left, which has been divided into five parts as indicated by the heavy dots. Establish at pleasure where the gores are to turn upward and place this point in its proper position as indicated by 6 in the side elevation in Fig. 1, from which drop a vertical line intersecting the top edge of the boat in plan, also at 6. Divide the curve 6 A into as many spaces as the boat is to contain gores, five in this case, as shown by the heavy dots 6 to 1. At pleasure locate a series of planes through the left half of the boat in the side elevation as shown by the planes 8 58, 7 59, 6 60, 6 61, 6 62 and 6 63, which will intersect the dotted curved line A *a*′ in elevation at *c d e f i* and *k* respectively.

From these small letters *c* to *k* in elevation, project perpendicular lines into the plan, cutting the dotted curved line A *a*, also at *c d e f i* and *k* respectively. In a similar manner project 7 and 8 in elevation into the plan, and obtain the points 7 and 8 as shown.

The next step is to obtain the proportional vertical views on the various planes 9 to A in elevation. As the planes 8 58, 7 59, and 6 60 in side elevation stand in a vertical position, the sections which will be obtained on these planes will show their true shape.

From the points 8, 7 and 6; *c*, *d* and *e*; 58, 59 and 60, draw horizontal lines to the left until they meet the center line E F as shown. Measuring from the

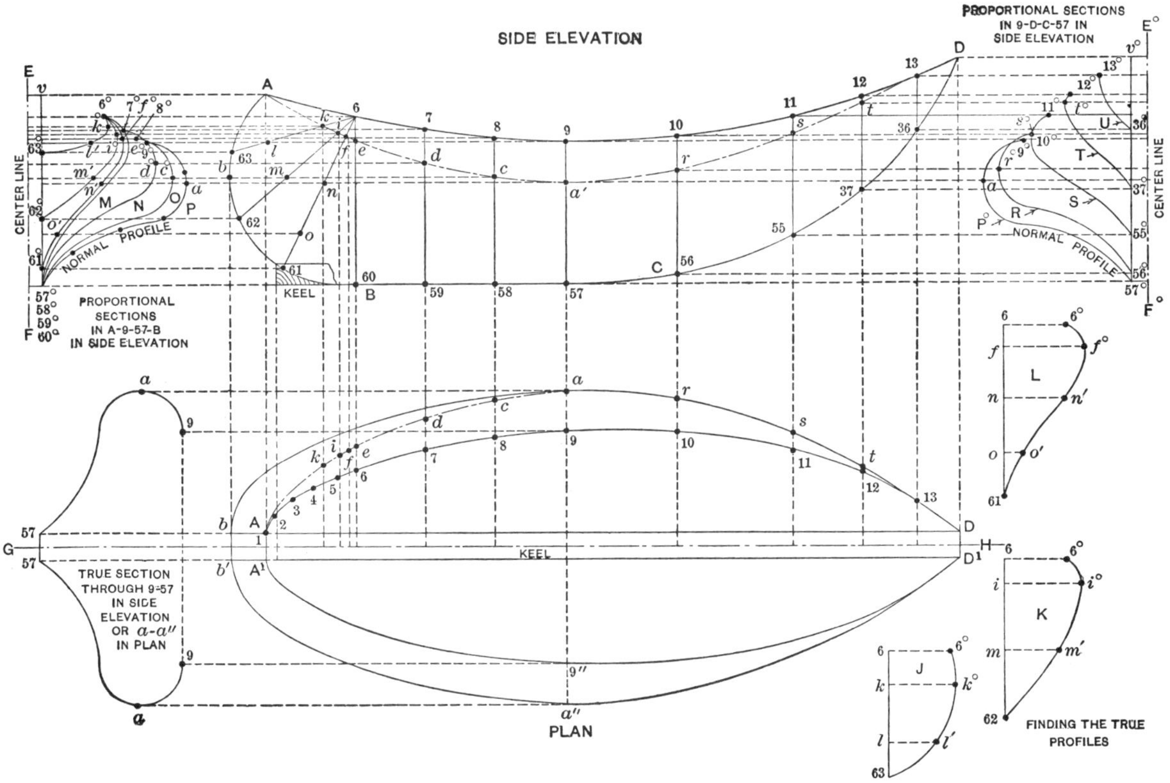

Fig. 1. Plan, Elevation and Method of Finding Proportional Vertical Section of Sheet Metal Boat

center line G H in plan take the projections to 8, 7 and 6; to *c d* and *e*, and place them on lines just drawn from similar letters and numbers in the side elevation, measuring in each instance from the line E F, thus obtaining the points of intersections 8°, 7° and 6°; *c*° *d*° and *e*° respectively. Through the points 8°, *c*°,

58° draw the profile O, giving it a proportional taper to the profile P. Through points 7°, *d*°, 59°, draw the profile N in proportion to the profile O. In a similar manner through the points 6°, *e*°, 60° draw the profile M, giving it a proportional taper to the profile N. These four profiles M, N, O, P, then represent the true sections on the planes drawn from 6, 7, 8 and 9 in the side elevation.

From the lower part of the planes at 61, 62 and 63 in the side elevation draw horizontal lines to the left until they intersect the keel line drawn from *v*, at 61°, 62° and 63°. From points *f*, *i* and *k* in the side elevation, draw horizontal lines to the left until they intersect the center line E F as shown. Measuring from the center line G H in plan take the distances to points *f*, *i* and *k* and place them on similar lettered lines just drawn from the side elevation, measuring from the line E F and obtaining the points *f*°, *i*° and *k*°. Through the points 6°, *f*°, and 61° draw a profile in proportion to the profile M, which will represent a vertical view on the plane 6 61 in the side elevation. In a similar manner through the points 6°, *i*°, 62° and 6°, *k*°, 63° draw graceful profiles in proportion to one another as shown. These last two profiles 6°, 62° and 6°, 63° represent the vertical views on the planes 6 62 and 6 63 in the side elevation.

The next step is to find the true sections on the planes 6 61, 6 62 and 6 63 in the side elevation, as follows: At pleasure establish any point as *l* on the plane 6 63; another at *m* on the plane 6 62, and two more at *n* and *o* on the plane 6 61. From these points *l*, *m*, *n* and *o*, draw horizontal lines to the left until they intersect the proper view, as indicated by *l*′, *m*′, *n*′ and *o*′. Take the various divisions on the planes in side elevation, as 6 *k l* 63, 6 *i m* 62, 6 *f n o* 61, and place them as shown by similar letters and numbers on the vertical lines to the right of the plan view, in the diagrams J, K and L respectively.

From the divisions on the vertical lines in J, K and L draw perpendicular lines indefinitely as shown. Measuring from the keel line *v* 57° in the sections to the left of the elevation, take the various distances to points 6° *f*° *n*′ *o*′ and place them on similar lines in diagram L measuring from the line 6 61, thus obtaining the points 6 *f*° *n*′ *o*′. Through the points trace a graceful profile as shown, which will represent the true section on 6 61 in the side elevation.

In a similar manner, measuring from the keel line *v* 57°, take the distances to points 6° *i*° *m*′, also to 6° *k*° *l*′, and place them on similar lettered and numbered lines in the diagrams K and J respectively, thus obtaining the points of intersection shown by similar letters and numbers. Draw the profile from 6° to 62 and from 6° to 63, which will represent respectively the true sections on the planes 6 62 and 6 63 in the side elevation.

These planes which are shown drawn through points 10, 11, 12 and 13 in the side elevation cannot be drawn at pleasure, but must be located in a manner which will be described as we proceed. To avoid a confusion of lines which would otherwise take place in so small a space, take a tracing of the side elevation of the boat A B C D, with the various planes 9 57, 8 58, 7 59, 6 60, 6 61, 6 62 and 6 63, and place it as shown by A^x B^x C^x D^x in the side elevation in Fig. 2, also reproducing the planes just mentioned as shown by similar numbers.

In a similar manner reproduce the half plan *b a* D A in Fig. 1 as shown by b^x a^x 14′ 1′ in Fig. 2, placing it in line with the elevation as shown. Take a tracing of the divisions from 1 to 9 in the plan in Fig. 1 and place them in a similar position in the half plan in Fig. 2, as shown from 1′ to 9′. Take tracings of the true profiles J, K, L, M, N, O, P, in Fig. 1, and place them as shown by J° K° L° M° N° O° and P° in Fig. 2.

As the boat in this case is to contain five gores, divide the sections J°, K° and L° each into five equal spaces, as shown from 6° to 63, 6° to 62 and 6° to 61 respectively. From these points perpendicular to the vertical lines, draw lines intersecting the vertical line at 25, 26, 46, 47 in J°; at 24, 27, 45, 48 in K°, and at 23, 28, 44 and 49 in L°. Take the various divisions on 6 63 in J°, 6 62 in K° and 6 61 in L° and place them on similar numbered planes in the side elevation, thus obtaining similar numbered intersections as shown by the heavy dots. From the intersections 1′ to 6′ in the half plan, project lines into the elevation, cutting the top line of the boat from 1 to 6 as shown.

In the true sections, P° represents the given profile, which is also divided into five equal spaces as shown, from which points lines are drawn perpendicular to 9 57, intersecting this line at 19, 32, 40 and 53. Take these divisions on 9 57 in P° and place them as shown by similar numbers on the plane 9 57 in the side elevation.

Through the various intersections on the planes 6 1, 6 63, 6 62, 6 61 and 9 57 in the side elevation, draw graceful free-hand curves, representing the longitudinal joints as follows: From 5 through 25, 24, 23, 19 to 14, which gives the elevation of A or the first gore; from 2 through 47, 48, 49, 53, then continue with a graceful curve until it meets the bow at 55, which gives the elevation of the fifth gore E. Divide this distance between 55 and 14 into as many parts as the boat is to have gores, minus the first and last gore. As the boat in this case is to have five gores, then the space between 55 and 14 is divided into three parts as shown by 36 and 37.

Complete the other two longitudinal joint lines by drawing graceful sweeps from 4 through 26, 27, 28, 32 to 36; from 3 through 46, 45, 44, 40 to 37. Thus the drawing of these last two joint lines gives the elevation of the second gore B, the third gore C and the fourth gore D. Where these joint lines intersect the bow of the boat at 36, 37 and 55, this gives the location from which the vertical planes are erected as 36 13, 37 12 and 55 11.

Establish another point between 11 and 9 as 10, and draw the vertical plane 10 56. By following the points of intersections in the side elevation, it will be found that they have been numbered starting at 1 to 14, to 25, to 26, to 36, to 37, to 46, to 47, to 55, to 56, to 63.

Where the longitudinal joint lines cut the planes 6 60, 7 59 and 8 58 at from 22 to 50, 21 to 51, and 20 to 52 respectively, take the various divisions on 6 60,

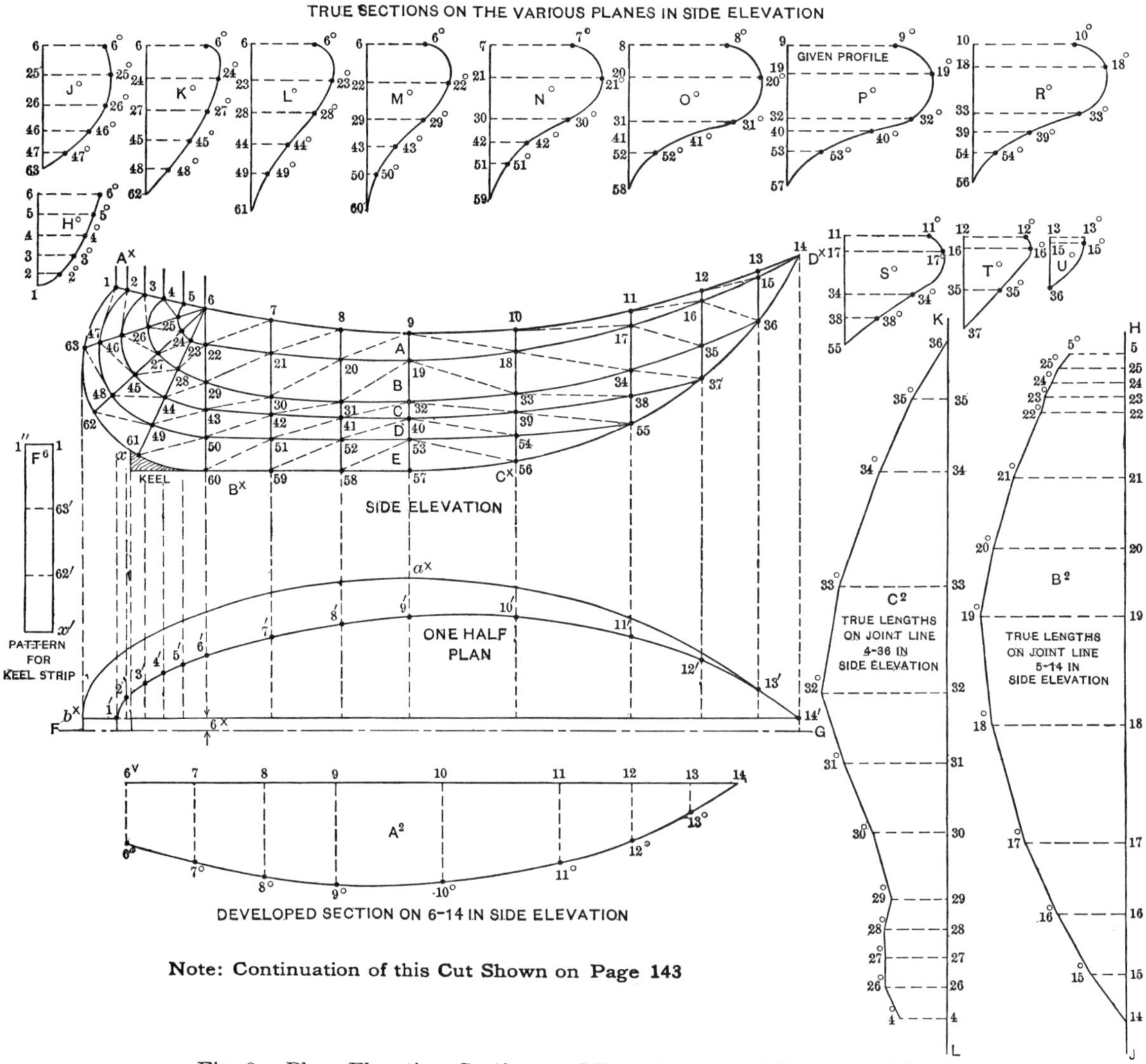

Note: Continuation of this Cut Shown on Page 143

Fig. 2. Plan, Elevation, Sections and True Lengths of Sheet Metal Boats

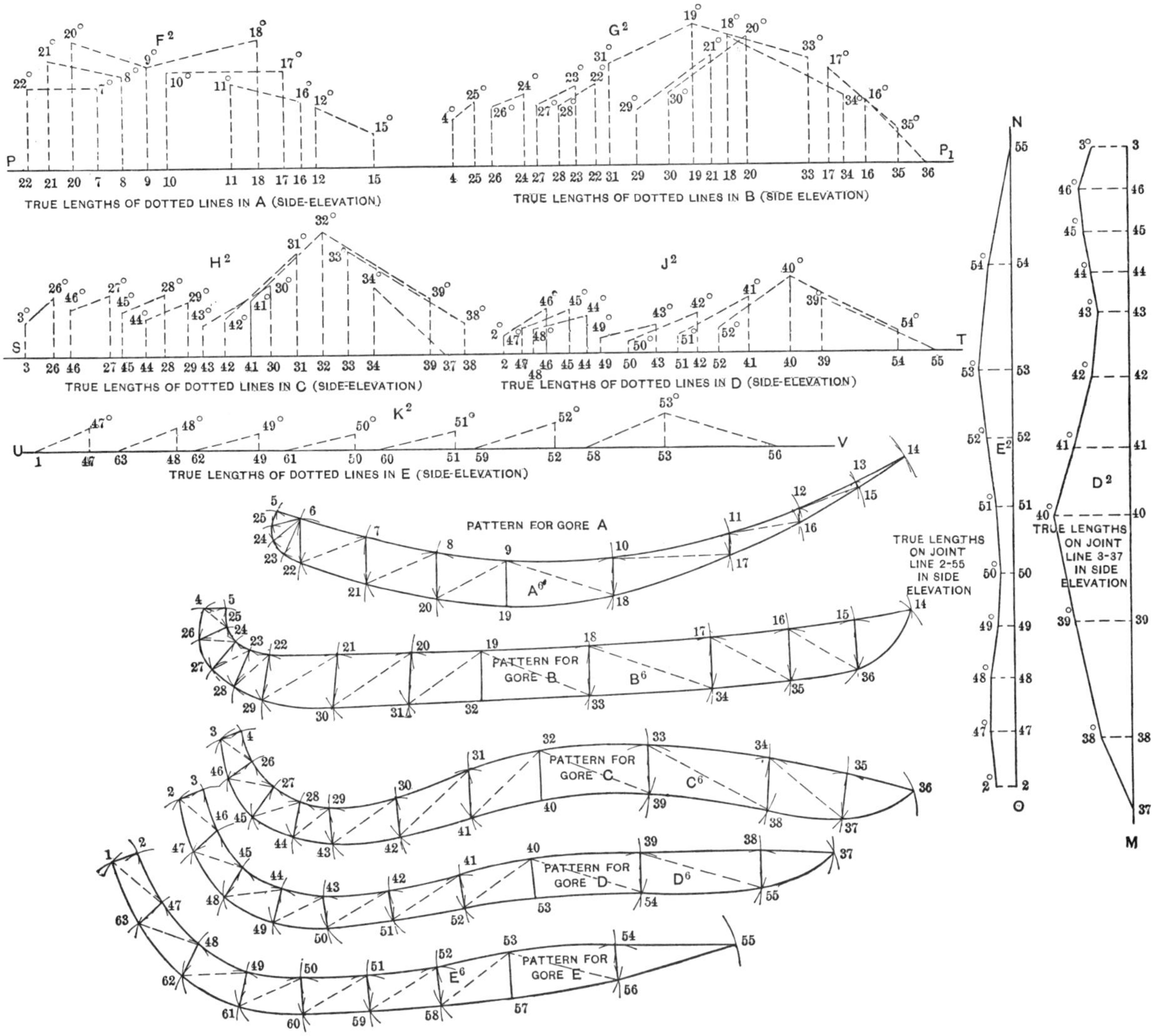

Fig. 2. True Lengths, Sections and Pattern Shapes of Sheet Metal Boats

7 59 and 8 58 and place them on similar numbered lines in the true sections as shown by similar numbered intersections on the vertical lines 6 60, 7 59 and 8 58 in the profiles M°, N° and O°. From these intersections on the vertical lines draw lines at right angles to 6 60, intersecting the profile M° at 22°, 29°, 43° and 50°, which divisions will give the proper girth when laying out the patterns. In a similar manner at right angles to the vertical lines in N° and O° from the various intersections on same, lines are erected until they cut the profiles as shown.

It now becomes necessary to find the true sections on the planes 10 56, 11 55, 12 37 and 13 36 in the side elevation as follows: Take a tracing of these various planes in 9 D^x 57 in the side elevation and place them as shown by similar numbers in the half elevation 9 D 57 in Fig. 1, noting where the planes cross the extreme outer edge of the boat at *r*, *s* and *t*. Extend the planes into the plan, so as to obtain similar intersections 10, 11, 12 and 13 for the top edge of the boat, and *r*, *s* and *t* for the outer extreme edge of the boat. The various intersections in elevation give the hights of the profiles to be found, while the various intersections in plan will give the projections.

To find the proportional true sections on the planes 10 56, 11 55, 12 37 and 13 36 proceed as follows: From the intersections 36, 37, 55 and 56 in the side elevation draw horizontal lines to the right, intersecting the keel line *v*° 57° at 36°, 37°, 55° and 56° respectively. In a similar manner from the intersections 10, *r*, 11, *s*, 12, *t* and 13 in the side elevation, draw horizontal lines until they intersect the center line E° F° as shown by similar numbers. Measuring from the line G H in plan take the various projections to points 10, *r*, 11, *s*, 12, *t* and 13 and place them on lines drawn from similar numbers in the elevation, measuring in each instance from the center line E° F°, and obtaining the points of intersections 10°, *r*°, 11°, *s*°, 12°, *t*° and 13°.

Through the intersections 10°, *r*°, 56° draw the profile R in tapering proportion to the profile P°. Through the points 11°, *s*°, 55° draw the profile S in proportion to the profile R. In a similar manner through points 12°, *t*°, 37° draw the profile T in proportion to S, and through points 13° 36° the profile U in proportion to T. Then will the profiles R, S, T and U represent the true sections through the planes 10 56, 11 55, 12 37 and 13 36 respectively in the side elevation.

Take a duplicate of the profile R, S, T and U and place them in Fig. 2 as shown by R°, S°, T° and U°. Now take the various intersections on the plane 10 56 in the side elevation and place them on the vertical line 10 56 in the true section R°. At right angles to 10 56 from the intersections 18, 33, 39, 54 draw lines cutting the profile as shown by similar numbers. In a similar manner transfer the intersections from the planes 11 55, 12 37 and 13 36 in the side elevation, on to similar numbered vertical lines in the true sections S°, T° and U° and obtain the intersections on the outlines of the profiles as shown.

Thus it will be noted that only the true sections J°, K°, L° and the given section or profile P° have been equally spaced; all the spaces on the balance of the true sections are unequal. This becomes necessary, so that graceful sweeps can be obtained along the longitudinal joints in the side elevation. Thus if all the true sections were divided into equal spaces, and these divisions projected to the vertical line in the section, thence to the corresponding plane in elevation, the joint line would become zig-zag, owing to the changing of each profile on the planes in elevation.

Connect by dotted lines the shortest opposite points in the side elevation as from 22 to 7, 21 to 8, 20 to 9, etc., as shown. These solid and dotted lines in elevation then represent the bases of sections which must now be constructed, whose altitudes are equal to the various hights shown by similar numbers in the various true sections. The true lengths on the various planes in elevation are shown by similar numbers in the true sections.

To find the true length of the upper edge line of the boat shown from 1 to 6 in the elevation, take the girth from 1 to 6 and place it as shown on the vertical line in H°. Through these small figures perpendicular to 6 1 draw lines making them equal in length to the distances measured from the line b^x 6^x in plan to points $1'$ to $6'$. A line traced through points thus obtained in H° as shown from 2° to 6° will be the true length of the upper line of the boat from 1 to 6 in elevation.

Next, to find the true length of the upper line 6 14, take this girth and place it on the horizontal line 6^v 14 below the plan as shown by similar numbers, through which at right angles to 6^v 14, draw lines indefinitely as shown. Measuring from the line 6^x $14'$ in plan take the various divisions to points $6'$ to $13'$ and place them on similar numbered lines in A^2, measuring in each instance from 6^v 14, thus obtaining the points of intersections 6° to 13°. A line traced through these points will give the true lengths along the upper edge of the boat 6 14 in elevation, and this developed section on 6 14 in the side elevation, will be known hereafter as A^2.

To obtain the true length of the joint line 5 14 in side elevation, take the various divisions 5 to 25, to 24, to 23, etc., up to 14 and place them on any vertical line as H J as shown by similar numbers. Through these numbers at right angles to H J draw lines as shown, making them equal in length to the various hights in the true sections having similar numbers, measuring always from the vertical lines in the various sections.

For example, to find the true length of 18 19 in the side elevation, place this distance on the line H J as shown from 18 to 19, from which points erect the perpendicular 18 18° and 19 19° equal respectively to the distances 18 18° in the true section R° and 19 19° in the true section P°. Then a line drawn from 18° to 19° in B^2 will be the true length desired. In this way, all of the true lengths in B^2 are obtained.

Take the various divisions on the joint line 4 36 in the side elevation and place them on the vertical line K L, from which points lines are erected equal in length to similar numbered hights in the true sections. As no two sections are numbered alike, no mistake can occur in finding the proper distance to be placed in C^2 in which the solid irregular line shows the true lengths of the various divisions required. In a similar manner take the girth of the various spaces in the joint lines 3 37 and 2 55 in the side elevation and place them on the vertical lines J M and N O respectively as shown by similar numbers. From these small figures, perpendiculars are drawn, equal to similar numbered distances in the proper true sections or profiles, thus giving the true lengths in diagrams D^2 and E^2.

In the side elevation, the various gores have been lettered A B C D and E. To find the true lengths of the dotted lines in A, take the various lengths of 7 22, 8 21, 9 20, etc., and place them on any line as P P_1 as shown by similar numbers in F^2, setting each inside of one another, to save space. From these small figures on P P_1 in F^2, erect perpendiculars equal to the various projections of similar numbered points in the true sections. Thus to find the true length of 8 21 in the gore A in side elevation, take this distance and set it off on the line P P_1 as shown by 8 21. From 8 and 21 erect the perpendiculars 8 8° and 21 21° equal to similar numbers in the true sections A^2 and N°. The distance from 8° to 21° in F^2 then gives the true length of the line 8 21 in gore A in side elevation.

In a similar manner take the lengths of the various dotted lines in gore B in side elevation, also placing them upon the line P P_1 and obtain the true lengths of these lines as indicated in diagram G^2. In precisely the same manner take the various lengths of the dotted lines in gores C, D and E in the side elevation and place them as shown by similar numbers on the lines S T and U V respectively, making the vertical hights in these three diagrams H^2, J^2 and K^2 equal to similar hights in the various true sections. As before mentioned, no mistake can occur in finding these true lengths as every point has a different number.

No true lengths need be found along the bottom of the boat 1 57 14, as this line shows its true length. The true sections having been found, also the various true lengths on all points shown in the side elevation, the various patterns are now in order, and we will begin with the gore A in the side elevation starting at 5 6.

Take the distance of 6° 5° in the true section H°, and place it as shown by 6 5 in the pattern shape A^6. Using 6 as center and 6° 25° in the section J°, 6° 24° in the section K°, 6° 23° in the section L° and 6° 22° in the section M° as radii, draw the various arcs in the pattern A^6 as shown by 25, 24, 23 and 22. Using 5 as center and 5° 25° in the true lengths in B^2 as radius, intersect the arc 25 in A^6 as shown. Again using the divisions in B^2 from 25° to 24°, 24° to 23° and 23° to 22°, set them off in the pattern A^6, as shown by the intersections 25 to 24, 24 to 23 and 23 to 22. Draw lines from 22, 23, 24 and 25 to 6.

With 6° 7° in the developed section A^2 as radius, and 6 in the pattern A^6 as center, describe the arc 7, which intersect by an arc struck from 22 as center, and 22° 7° in the true lengths in F^2 as radius. Now with 22° 21° in the diagram B^2 as radius, and 22 in the pattern A^6 as center, describe the arc 21, which intersect by an arc, struck from 7 as center, and 7° 21° in the true section N° as radius. Proceed in this manner, using alternately first the proper division in A^2, then the true length in the diagram F^2; the proper division in B^2, then the proper length shown along the profile in the true section, until the intersection 14 in the pattern A^6 has been obtained, which completes the pattern for the gore A in side elevation.

The pattern for gore B is shown by B^6. The spaces along 5 14 are obtained from the spaces along 5 14 in the lower part of the pattern A^6. The length of the dotted lines in B^6 are obtained from the true lengths in G^2, while the length of the solid lines in B^6 are obtained from the curved parts of the various true sections of similar numbers. The divisions along the lower edge in the pattern B^6 are taken from the true lengths in C^2, excepting the distance 36 14 in the pattern B^6, which is taken from 36 14 in the side elevation which shows its true length.

The pattern for the gore C is shown by C^6. The divisions along the upper edge are taken from the lower edge of B^6, the divisions along the lower edge in C^6 are obtained from the true lengths in D^2, excepting 36 37 in C^6, which is taken from the side elevation. The length of the solid and dotted lines in C^6 are taken respectively from the curved profiles in the true sections and the true lengths in diagram H^2. So in D^6, which represents the pattern for the gore D, the divisions along the upper edge line are obtained from the divisions in the lower edge line in pattern C^6. The divisions along the lower edge in D^6, being obtained from the true lengths in E^2, excepting the division 37 55 in D^6, which is obtained from similar numbers in the side elevation. The lengths of the solid and dotted lines in the pattern D^6 are taken respectively from the divisions in the curved profiles in the true sections having similar numbers and the true lengths in the diagram J^2.

The last pattern against the keel, shown in side elevation by E, is shown developed by E^6. The divisions along the upper edge line are obtained from the divisions in the lower edge line in the pattern D^6 while the divisions along the lower edge line of the pattern E^6 are obtained from the lower line of the boat in the side elevation which shows its true length along the keel, as shown by similar numbers. The lengths of the various solid and dotted lines in the pattern E^6 are obtained respectively from the proper numbered divisions in the curved profiles in the true sections and the slant lines in diagram K^2.

The pattern for the keel strip, following the continuation of the wooden keel, is simply a rectangular piece of metal shown in the diagram F^6 and is developed as follows: Take the girth from 1 to 63 to 62 to x in the side elevation, and place it as shown by similar numbers in F^6, making the width I 1″ as wide as the width of the wooden keel, or twice the width of 6^x in the one-half plan. This completes all of the patterns, to which laps must be allowed for nailing to the wooden keel, as well as for riveting the overlapping longitudinal joints.

The various gores must be raised or stretched to conform to the various true sections shown. When raising the various gores, the various true full-size sections should be cut out, flanges allowed and nailed on the proper plane line in the full-size side elevation placed on the floor. This then gives a model, after which the various gores can be fitted and then riveted and the joints and rivets sweated with solder. The seats at each end of the boat should be made up in the form of an air-tight compartment, to keep the boat from sinking if overturned.